Studies in Computational Intelligence

Volume 982

The series “Studies in Computational Intelligence” (SCI) publishes new developments and advances in the various areas of computational intelligence—quickly and with a high quality. The intent is to cover the theory, applications, and design methods of computational intelligence, as embedded in the fields of engineering, computer science, physics and life sciences, as well as the methodologies behind them. The series contains monographs, lecture notes and edited volumes in computational intelligence spanning the areas of neural networks, connectionist systems, genetic algorithms, evolutionary computation, artificial intelligence, cellular automata, self-organizing systems, soft computing, fuzzy systems, and hybrid intelligent systems. Of particular value to both the contributors and the readership are the short publication timeframe and the world-wide distribution, which enable both wide and rapid dissemination of research output.

Indexed by SCOPUS, DBLP, WTI Frankfurt eG, zbMATH, SCImago.

All books published in the series are submitted for consideration in Web of Science.

More information about this series at http://www.springer.com/series/7092

Monica Malvezzi · Dimitris Alimisis ·
Michele Moro
Editors

Education in & with Robotics to Foster 21st-Century Skills

Proceedings of EDUROBOTICS 2020

Springer

Editors
Monica Malvezzi
Department of Information Engineering
and Mathematics
University of Siena
Siena, Italy

Dimitris Alimisis
European Lab for Educational Technology
Edumotiva
Sparta, Greece

Michele Moro
Department of Information Engineering
Università degli Studi di Padova
Padova, Italy

ISSN 1860-949X ISSN 1860-9503 (electronic)
Studies in Computational Intelligence
ISBN 978-3-030-77021-1 ISBN 978-3-030-77022-8 (eBook)
https://doi.org/10.1007/978-3-030-77022-8

This Springer imprint is published by the registered company Springer Nature Switzerland AG
The registered company address is: Gewerbestrasse 11, 6330 Cham, Switzerland

Preface

This book collects the papers presented at the 2020 edition of Educational Robotics International Conference (EDUROBOTICS). The conference was originally planned for November 2020 in Siena, Italy, but due to COVID-19 restrictions, it was transformed into an online event and postponed to February 2021. Over the years, EDUROBOTICS conference contributed to building a community of researchers and educators in Educational Robotics at European and international level. The conference was previously organized in Venice (2008), Darmstadt (2010), Riva del Garda (2012), Padova (2014), Athens (2016) and Rome (2018). Notwithstanding the online modality and the uncertainties related to the particular situation, 30 papers were submitted to the conference, all the papers were reviewed by at least two expert reviewers, and 19 of them were finally selected for the presentation at the conference and included in this book.

The conference started with the keynote talk by Paulo Birkstein and Arnan Sipitakiat. Paulo Blikstein is Associate Professor of Education and Affiliate Associate Professor of Computer Science at Columbia University, where he directs the Transformative Learning Technologies Laboratory. Arnan (Roger) Sipitakiat directs the Teaching and Learning Innovation Center (TLIC) at Chiang Mai University in Thailand. A Faculty and Researcher at the Computer Engineering Department, he also directs the Learning Inventions Laboratory (LIL).

The keynote speakers introduced the history of educational robotics, discussed research-based design principles, presented examples of good and bad design and demonstrated a brand-new lineage of open-source platforms for educational robotics that will help lead the field into the future. The conference ended with an invited talk by Gary Stager, one of the world's leading experts and advocates for computer programming, robotics and learning-by-doing in classrooms. Dr. Stager presented an expansive view of robotics to serve a diverse population of learners, discussed contexts for using robotics as a vehicle for knowledge construction and shared models for inspiring teachers to embrace the potential of robotics as an expressive medium and incubator of powerful ideas.

The chapters are organized in this volume in four sections.

The first section includes the papers that report new developments in educational robotics. This branch of robotics is attracting more and more attention, and several interesting solutions are present in the literature. The seven chapters in this section present experiences with educational robots in training students' problem-solving skills (Chapter 1) and talents (Chapter 2), solutions to continue education in robotics also under COVID-19 situation (Chapters 3 and 4), multi-use robots for technology education (Chapter 5), debugging strategies in robotic simulation environment (Chapter 6) and AI robots and AI literacy (Chapter 7).

The second section presents three interesting robot designs for social applications including children with ASD (Chapter 8), children in pediatric protective isolation units (Chapter 9) and elderlies (Chapter 10).

The third section includes papers related to the Inclusive Robotics for a Better Society (INBOTS) project, funded by the European Commission and aimed at fostering a multidisciplinary discussion on the role of robots in the overall society, including technical and non-technical aspects. In this section, a review on the application of virtual reality (VR) and augmented reality (AR) techniques for education and training (Chapter 11), current trends and shortcomings on educational robotics curricula (Chapter 12), a dance-based STEAM education project with the tool ENGINO (Chapter 13) and an example of curriculum for the robotics education of assistant nurses in Sweden (Chapter 14) are presented.

The fourth section presents some works dealing with new studies and methodological issues in educational robotics: a study on the educational robotics acceptance by teachers (Chapter 15), a project aiming at building a community on educational robotics (Chapter 16), an analysis leveraging on the "trial and error" strategy (Chapter 17), an educational robotic project involving students from school to university (Chapter 18) and work on the use of educational robotics in primary schools and its place in the curricula (Chapter 19).

We thank the authors for their contribution to this book. Special thanks go to the EDUROBOTICS 2020 program committee members who have reviewed the papers and provided important help to authors to improve their manuscripts.

March 2021

Monica Malvezzi
Dimitris Alimisis
Michele Moro

Organization

Program Committee Members

Dimitris Alimisis	Edumotiva, Greece
Richard Balogh	Slovak University of Technology in Bratislava, Slovak Republic
Ansgar Bredenfeld	Dr. Bredenfeld UG, Germany
Dave Catlin	Valiant Technology, UK
Linda Daniela	University of Latvia, Latvia
Edoardo Datteri	Università degli Studi di Milano-Bicocca, Italy
Amy Eguchi	University of California San Diego, US
Nikleia Eteokleous	Frederick University, Cyprus
Paolo Fiorini	University of Verona, Italy
Reinhard Gerndt	Ostfalia University of Applied Sciences, Germany
Ilkka Jormanainen	University of Eastern Finland, Finland
Ken Kahn	University of Oxford, UK
Martin Kandlhofer	Graz University of Technology, Institute for Software Technology, Austria
Chronis Kynigos	University of Athens, Greece
Wilfried Lepuschitz	Practical Robotics Institute Austria
Monica Malvezzi	University of Siena, Italy
Emanuele Micheli	Scuola di Robotica, Italy
Michele Moro	University of Padua, Italy
Britt Östlund	KTH, Sweden
Evangelos G. Papadopoulos	National Technical University of Athens, Greece
Alfredo Pina	Public University of Navarra, Spain
Maximiliano Romero	Università IUAV di Venezia, Italy
Theodosios Sapounidis	International Hellenic University, Greece

Sandra Schön	Graz University of Technology, Austria
Igor Verner	Technion - Israel Institute of Technology, Israel
Anton Yudin	Moscow State Technical University, Moscow, Russian Federation, Russia

Contents

New Developments in Educational Robotics

Fostering Students' Problem-Solving Skills Through Educational Robotics in Primary School

Francesca Gratani, Lorella Giannandrea(✉), Alessandra Renieri, and Martina Annessi

Department of Education, Cultural Heritage and Tourism, University of Macerata, 62100 Macerata, MC, Italy
{f.gratani,lorella.giannandrea,alessandra.renieri}@unimc.it

Abstract. Problem-solving is considered as one of the key competencies for the 21st century learners. In recent years, Educational Robotics (ER) proved to be a powerful tool to promote both computational thinking and problem-solving ability. However, research is still lacking in pathways to integrate ER into lower education curricula in order to use it as a real 'cognitive artifact'. Furthermore, ER and coding are often introduced at school through short-term extra-curricular activities. This research paper reports on an extended path on both coding and ER, involving an Italian fourth-grade class which was fully integrated into the class curriculum. The contribution is thus aimed at analyzing the effects of ER integration on problem-solving skills development, focusing on three main components of this ability (reported on Lucangeli's SPM test): comprehension, representation, and categorization. Multiple comparisons analysis was carried out using Friedman and Nemenyi post-hoc tests. The results show an improvement in students' problem-solving skills. The strong points outlined in this path are its interdisciplinary and multimodal approach. The various activities proposed led students to regularly argue, negotiate, and collaborate, working on multiple domains of knowledge and with different kinds of tools. Moreover, the design of the path enhances the alternation of some binomials: theory-practice, frontal lesson-cooperative learning, and analog-digital dimension.

Keywords: Educational robotics · Primary school · Problem-solving skills

1 Introduction

Today, students of any grade need new tools and mental structures to 'read' and decode the increasingly complex reality. Therefore, a growing attention is paid to learning processes and to Computational Thinking (CT), considered as a new essential skill for every child's

The article is the result of a common vision among the authors with the following responsibilities: Francesca Gratani is the author of Sections 2, 3.1, and 3.3; Lorella Giannandrea is the author of Sections 1 and 5; Alessandra Renieri is the author of Section 4; Martina Annessi is the author of Section 3.2.

M. Malvezzi et al. (Eds.): EDUROBOTICS 2021, SCI 982, pp. 3–14, 2021.
https://doi.org/10.1007/978-3-030-77022-8_1

learning path. In fact, the ability to think computationally is strongly correlated to the ability to logically solve problems in different technology-enhanced contexts [1].

Norman [2] referred to cognitive artifacts as artificial devices designed by man to act on information, and especially to preserve, present, and operate on it. Then, these artifacts have the potential to amplify and strengthen human minds capacities, guiding their mental activity and influencing their execution of a task.

In the last few years, Educational Robotics (ER) has been acknowledged similar potentialities. Besides, fostering many transversal skills, ER activities can support CT and problem-solving processes, stimulating students to identify action sequences, to generate and test hypotheses, and to verify them on something tangible and contextualized. Students deal with the feedback resulting from the robot/environment interaction [3], which leads them to revise, re-plan, and reflect on what they have done.

Studies also stress that ER effectiveness is highly influenced by the way teachers propose it to their pupils [4, 5]. The use of different methods may have an impact on the student performance, especially considering CT skills [6]. However, the research is still lacking pathways to integrate ER into lower education curricula in order to use it as a real 'cognitive artifact'.

In our study, we introduced primary school students to CT and problem-solving through an extended and interdisciplinary path both on coding and ER. This path, fully integrated into the class curriculum, was enriched by a multimodal approach, which include both plugged and unplugged phases. The contribution is thus aimed at analysing the effects of ER integration on problem-solving skills development. In particular, we focused on the evolution of three main components of this ability: comprehension, representation, and categorization.

Finally, the paper has the following structure: Sect. 2 provides a brief theoretical background on ER and problem-solving processes; Sect. 3 describes aims, methods, context/materials, and the experimental path; Sect. 4 reports the study results and their discussion; the last section outlines our conclusions and final considerations.

2 Problem-Solving and Robotics

2.1 State of the Art

Recent technological advances require educational institutions to reform and enhance the traditional school curriculum, in order to foster CT skills and so-called 21st-century skills. Many recent studies recognize the potential of ER [7–12]. ER is considered as an innovative model, not only to teach and learn programming, but also to empower transversal skills, such as problem-solving [13, 14], metacognition [3, 15], critical thinking [10], creativity [16], and collaboration [17]. Moreover, it can promote interdisciplinarity, and engage students in an exciting and playful learning environment [11]. The project-based approach, supported by robotics, increases students' involvement and motivation, affecting positively their performance [18].

However, pupils should have the possibility to actively create their robotic artefacts, understanding what a robot is and how it works, to really benefit from ER [5, 19]. In the learning process, they often deal with abstract concepts and processes. According to Ersoy et al. [20], this over abstraction can be overcome by the programming of educational robot sets. ER gives indeed students the opportunities to practice and convey the concepts studied in class into reality, in order to better understand them [21]. Working with programmable robots brings children to work in the real context, marked by all its variables and uncertainties [3].

A recent study by La Paglia et al. [3] evaluates the effects of a robotics laboratory on transversal high-level cognitive functions in children. The results prove that educational systems based on robotics encourage the use of specific cognitive and attentive abilities, strengthening mental processes, and having repercussions on executive functions. In fact, constructing and programming a robot to make it achieve a goal, requires many abilities related to CT, such as mental anticipation, logical and procedural thinking, pattern recognition, design of algorithms, and debugging [3, 22, 23]. Therefore, at a methodological level, ER proves to be a powerful tool to promote problem-solving and problem posing processes [24].

Varnado [25] referred to problem-solving as an inquiry process intentionally implemented by children to pose, examine, and find solutions to scientific problems, through four interactive phases: (1) identifying and defining the problem; (2) researching and analysing relevant information; (3) generating and implementing solutions to the problem; (4) evaluating and revising the best possible solution.

However, problem-solving strategies can vary in their degree of structuring [24]. Well-structured problems tend to be static and simple, while unstructured problems tend to be complex and dynamic [26]. In pre-school and primary school, studies suggest starting with simple and well-structured challenges that help children to create mental models useful for the resolution phase [27]. Then, children can move on to more complex, open, and dynamic problems. Through this iterated dynamic, they will finally develop a flexible and strategic attitude of analysis and resolution [24], as well as forecasting skills [28]. Realizing 'what' and 'how' they have learned leads children to internalize, communicate, correlate, and transfer these understandings to other daily problems [14].

Teachers can support this process by adopting coaching strategies. Rather than telling children what to do, they could observe them and ask guiding questions, facilitating both articulation and reflection. This will make it easier for teachers to find out the students' strengths, while enhancing their independent learning and problemsolving abilities [4].

3 Methodology

3.1 Aims and Scopes

This research aims at analysing the use of robotics in primary school in order to develop problem-solving skills in children. The experimental path is therefore oriented to initiate pupils to CT through two working phases (unplugged and plugged) on coding and robotics, to improve pupils' problem-solving ability, and to evaluate it in the long run.

The unplugged phase introduces the first activities, based on Coding. This was experimented at different levels (graphic and motor) without the aid of technology.

Then, the plugged phase foresees the experimentation of some robotics tools, such as the Bee-Bot and the LEGO We-Do 2.0 kit. These tools were used to program paths and to build artifacts which had "The Iliad", the famous poem by Homer, as an integrating background-theme.

A peculiar feature of this didactic project can be identified in its interdisciplinary perspective. In particular, the design system involves the following subjects:

- Italian: the subject is integrated through a path of narrative reading and guided understanding of the poem;
- History: the subject facilitates temporal orientation through the comparison between various civilizations (e. g. Greeks and Romans) and the creation of timelines on which to program the robot's path;
- Geography: the subject initiates spatial orientation in European maps, using the latitude and longitude coordinates to locate the city of Troy;
- Art: the subject is functional to carry out Pixel Art activities that provide iconic representations of the poem by means of letter-number coordination (e.g. the Trojan horse);
- Technology: the subject allows to use PC programs in order to document the activities carried out in a collaborative way (Padlet noticeboard).

Moreover, the connective cut of the path enhances the alternation of some binomials: theory-practice, frontal lesson-cooperative learning, and analog-digital dimension. The project is finally based on the awareness of achieving interdisciplinary and constructive learning, starting from a peer negotiation of meanings.

3.2 Context and Materials

The longitudinal experimental project took place at the Omni-comprehensive Institute of Amandola, located at the foot of the Sibillini Mountains, in Marche region. It involved a fourth-grade class and lasted about a year, from the second term of the fourth year until the first term of the fifth year. The class was composed of 20 pupils: 5 males and 15 females, including 1 foreign pupil and 2 with special educational needs.

The evolution of the students' problem-solving ability had been detected through the "SPM-Ability to solve mathematical problems" test, developed by Daniela Lucangeli and colleagues [29]. This test consists of four problems which are decomposed into the "five components that were shown to explain most of the total variance related to mathematical problem-solving ability" [29]. The ability to solve problems was indeed examined through the following five dimensions (or skills):

- comprehension of the information contained in the problem and its relationships;
- representation of information through a scheme that structures and integrates it;
- categorization of the problem according to its structure;
- planning the solution procedure;
- assessment of the correctness of the implemented procedure.

The test was administered four times, at the beginning and at the end of each phase. Each administration involves understanding and solving a problem of increasing difficulty. For the first three skills, students have to answer multiple-choice questions, choosing one of the four proposed alternatives:

- irrelevant answer: it reports information that is present in the text of the problem, but is not needed for the solution;
- incorrect answer: it reports information that, if used, leads to an incorrect result;
- partially correct answer: it reports correct but not complete data for the solution;
- correct answer: it reports all the useful data for the solution [29].

Thus, it is important that students read the problem carefully and begun to solve it after having performed the preliminary steps and marked the correct answers for each dimension. Only at this point could they solve the problem and self-assess the correct execution of the procedure. The scheduled time for performing the test is one hour, but it can be extended to one hour and thirty minutes maximum. Before administering the test, it was necessary to explain its structure, showing the test protocol, and verifying the students' understanding of the instructions.

As for the learning strategies and tools, in the first phase we used methodologies aimed at a first conceptualization of the term 'Coding'. In particular, we adopted some synergic methodologies, such as the brainstorming, which is linked to cooperative learning and peer tutoring strategy. That methodology was thus essential to face the challenges of graphic and motor Coding. These working sessions were characterised by the absence of robotic aids. The focus of this phase was indeed to enhance the pupils' naive ideas. The maps written on the board brought out the ability to problematize, while the worksheets on graphic Coding supported the programming and the narration of a path. They also fostered the ability to follow the instructions in Pixel Art activities and the body execution of the path in the class grid.

For the second phase we proposed the alternation of two methodologies: frontal lesson and cooperative learning. Finally, this phase offered, as an added value, the possibility to use various robotic tools, such as the Bee-bot and the LEGO We-Do 2.0 kit.

3.3 The Experimental Path

As mentioned before, the design of the entire path is divided into two phases: unplugged and plugged. Every phase is then divided into six working sessions (see Table 1), following the same structure.

The first six didactic sessions were designed in order to facilitate a first familiarization with the concept of Coding. The initial step foresaw the reading of a letter sent by a bee containing the challenge to be faced, and a brainstorming session. The central step included some multimodal Coding activities: completion of worksheets to program paths in a grid and carry out Pixel Art activities; body execution of the paths in the class grid. The final step was based on the transcription of a "Codinglossary". For each term learnt, pupils 'won' a symbol to stick on a poster disclosing a secret code to enter the "World of Robots".

The following six sessions were structured in order to combine ER with the curricular didactics in an interdisciplinary way. The integrating background-theme of this phase is the famous poem by Homer, "The Iliad". The reading of a part of the poem was set forth as the initial moment. It was accompanied by the analysis of the text through maps and bullet lists on the interactive whiteboard. The narrative dimension was combined with the digital one, since the theoretical part of the story was tackled in a practical way through Coding plugged activities. The pupils were divided into groups of five members and involved in various challenges related to the poems' characters and stages, such as:

- reaching the city of Athens in the map-grid in the shortest time possible;
- reaching one of the civilizations on the timeline-grid previously created;
- recomposing the characters' names by moving on a poster-grid showing the scrambled letters;
- reconstructing the main characters' family tree.

Each group solved these challenges creating a resolution path were the Bee Bot was subsequently programmed. The groups were also involved in the collective creation of a digital comic strip about the poem. In the last sessions, we proposed a debate between groups. Each one had to support of one of the two population (Achaeans or Trojans), and decide on the strong points of their choice. Then, the group leaders explained the reasons of each group, trying to persuade the other team. After the final reading of the poem and the election of the winning faction, we invited pupils to build their own robots, using the LEGO We-Do 2.0 kit. They indeed created and programmed the "Trojan horse" according to the instructions provided by the tablet app. Finally, their final considerations on the experience were collected and shared.

Table 1. The experimental path.

Unplugged phase: 6 working sessions	
Aim	Sessions' activities
To familiarize with concept of Coding	1. Pixel Art 2. Graphic Coding (grid on a sheet) 3. Graphic Coding (grid on a sheet) 4. Motor Coding (grid on the classroom floor) 5. "Cody Word" (grid as a crossword puzzle) 6. Coding activity with colours
Plugged phase: 6 working sessions	
Aim	Sessions' activities
To combine ER with the curricular didactics in an interdisciplinary way	1. Programming Bee-bot both on the map-grid and on the timeline-grid 2. Programming Bee-bot on a poster grid showing scrambled letters 3. Programming Bee-bot on a poster showing the characters' family tree 4. Debate and Pixel Art 5. Creating a digital comic strip about the poem 6. Building the "Trojan Horse" through Lego We-Do 2.0 kit

4 Results

The SPM test's questions aimed at highlighting the dimensions described above and the level of competence shown for each of them.

In order to assign the right score to the answers given, the test provided the teacher with correction grids. Thus, we assigned a score from 1 to 4 for every dimension, according to the selected answer (1 - irrelevant answer or omission, 2 - incorrect answer, 3 - partially correct answer, 4 - correct answer).

Only the first three dimensions were explored through multiple-choice questions. The score for the planning and the assessment referred to the correctness of the procedure implemented and was, therefore, assigned by the teacher in a subjective/open-ended manner.

In this research, we chose to examine only the first three dimensions (comprehension, representation, categorization), and to analyse only the data from multiple answer items. In particular, we traced the evolution of each skill in a diachronic way, parallel to the development of the project.

The statistical analysis was carried out using the free software R (v. 4.0.3 2020.10.10). The following administrations showed a gradual evolution of the skills observed, as can be seen in Table 2 and in Fig. 1.

Table 2. Test administrations results.

Skill	First test		Second test		Third test		Fourth test	
	Mean	Standard Deviation	Mean	Standard Deviation	Mean	Standard Deviation	Mean	Standard Deviation
Comprehension	3,20	0,77	3,55	0,60	3,75	0,55	3,95	0,22
Representation	2,95	0,89	3,25	0,85	3,20	0,77	3,90	0,31
Categorization	2,60	1,35	2,85	1,34	2,75	1,37	3,95	0,22

We tested normality for each population (for each skill and for each test) using the Shapiro-Wilk test and we found out that none of them was normally distributed ($\alpha = 0.01$) as can be seen in Table 3.

Then we performed multiple comparisons for assessing the statistical significance of the improvement of the examined dimensions: we did not use ANOVA since its assumptions were violated. Thus, we used Friedman test and Nemenyi post-hoc test.

According to the Friedman p-values, shown in Table 4, there is a significant statistical difference in the tests of the three skills ($\alpha = 0.01$). Given this condition we performed the Nemenyi post-hoc ($\alpha = 0.05$), in order to known where the differences took place: in Comprehension Test 1–Test 4 ($p = 0{,}014$), in Representation Test 1–Test 4 ($p = 0{,}014$), in Categorization Test 1–Test 4 ($p = 0{,}0079$) and Test 3–Test 4 ($p = 0{,}355$).

In our opinion the improvement of these problem-solving dimensions was connected to the 'narrative' aspects, such as the understanding of the text and the ability to represent. This can be explained by referring to the type of activities that had been proposed.

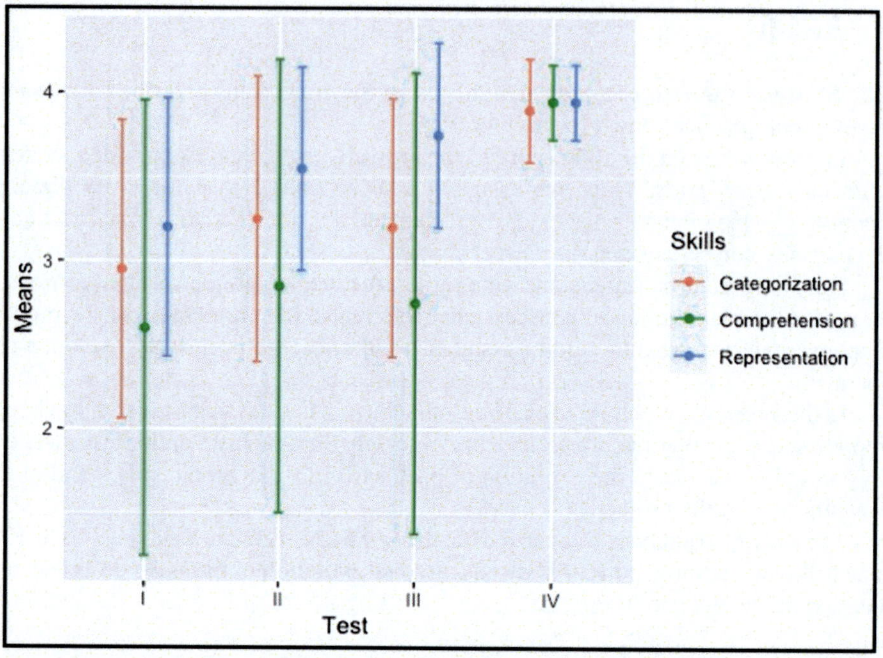

Fig. 1. Mean and standard deviation of the three skills for each test. Representation of the evolution for the mean (point) and standard deviation (error bar) in each performed test.

Table 3. Shapiro-Wilk test - p value.

Skill	First test	Second test	Third test	Fourth test
Comprehension	0,0002911	4,023e−05	5,056e−07	2,66923e−09
Representation	0,0003379	0,0001874	0,0002911	1,846e−08
Categorization	0,01047	0,001158	0,002617	2.693e−09

In fact, the tasks of designing and using robots were supported through narrative and collaborative activities, based on the debate and interpretation of stories.

As described in the previous section, each construction and programming activity was introduced by a narrative motivational phase (presentation of a stimulus, brainstorming, and discussion) and followed by a debriefing moment. On this occasion, the students reflected on the work carried out and consolidated new ideas by discussing and documenting what had been achieved. Therefore, besides the exercise of coding and robotics skills, students were involved in anticipation, storytelling and the planning activities, developing the habit of arguing and presenting their products to the group. The construction of new knowledge through peer negotiation is certainly one of the added values achieved in this path.

Table 4. Friedman test and Nemenyi post-hoc test.

Friedman p-values		*0,0002999*			*0,0002953*			*0,0006836*		
Nemenyi p-values		Test 1	Test 2	Test 3	Test 1	Test 2	Test 3	Test 1	Test 2	Test 3
	Test 2	0,650			0,827			0,8831		
	Test 3	0,122	0,725		0,856	1		0,9614	0,9948	
	Test 4	*0,014*	0,256	0,856	*0,014*	0,140	0,122	*0,0079*	0,0681	*0,0355*

The documentation of the path was reported in a virtual Padlet noticeboard, called "*Iliade in Coding*" (https://it.padlet.com/amartina94/74spfpvan42p).

Finally, from the final considerations of the students, a general appreciation of the project came out. In particular, they grasped and appreciated its interdisciplinary nature and the stimulus provided to metacognitive and problem-solving skills. We report some of these considerations:

- "The educational robotics course was interesting because it taught me how to reason and to make interdisciplinary connections";
- "Through robotics we learned to reason";
- "I learned that no knowledge walks alone";
- "This course was interesting as a mix of subjects";
- "I had the opportunity to deepen the historical-geographical subjects through Bee-bot in the grid".

5 Conclusions

One of the goals of ER is to allow students to construct their own models and representations, moving from the concrete to the abstract, trying to improve problemsolving by manipulating objects.

The experimentation process presented in this paper tries to construct a learning path through the alternation of different postures: merging theory and practice, proposing frontal lessons and cooperative learning activities, using narrative and storytelling to promote the development of digital literacy.

We recognize as strong points of the outlined path its interdisciplinarity, its multimodal approach, and the constructive learning promoted. The various activities proposed led students to regularly argue, negotiate, and collaborate, working on multiple domains of knowledge and with different kinds of tools. Primary students were engaged in collaborative processes, and prompted to reflect on their artifacts, using and creating metacognitive strategies. This had an impact when developing intellectual autonomy.

The results collected in this study seem to be promising. Indeed, ER could be a powerful tool for promoting problem solving skills in a learning approach where:

- students were involved in the design of meaningful projects and develop their ideas in autonomous ways;
- they constructed their knowledge both in the real word (making their artifacts) and in the digital environment (coding and digital documentation of the project);
- they experienced cognitive conflicts and received intrinsic feedback [30], especially when robots actions did not correspond to the expected ones;
- they learned to reflect on the artifacts and the projects they realized;
- they learned to negotiate and discuss their argumentation, developing social skills useful for their future life.

The good results obtained, despite the limitations of the research due to the low number of subjects involved, encourage to move forward in the research on the use of ER for the curricular development of subjects in primary schools.

References

1. Wing, J.M.: Computational thinking. Commun. ACM **49**(3), 33–35 (2006). https://doi.org/10.1145/1118178.1118215
2. Norman, D.: Things that Make us Smart: Defending Human Attributes in the Age of the Machine. Diversion Books, New York (1994)
3. La Paglia, F., Francomano, M., Riva, G., La Barbera, D.: Educational robotics to develop executive functions visual spatial abilities, planning and problem-solving. Ann. Rev. CyberTherapy Telemed. **16**, 80–86 (2018)
4. Chen, X.: How does participation in FIRST LEGO league robotics competition impact children's problem-solving process? In: Robotics in Education. RiE 2018. Advances in Intelligent Systems and Computing, vol. 829. Springer, Cham (2018). https://doi.org/10.1007/978-3-319-97085-1_16
5. Alimisis, D.: Educational robotics: open questions and new challenges. Themes Sci. Technol. Educ. **6**(1), 63–71 (2013)
6. Atmatzidou, S., Demetriadis, S.: Advancing students' computational thinking skills through educational robotics: a study on age and gender relevant differences. Robot. Auton. Syst. **75**, 661–670 (2016). https://doi.org/10.1016/j.robot.2015.10.008
7. Juškevičienė, A., Dagienė, V.: Computational thinking relationship with digital competence. Inform. Educ. **17**(2), 265–284 (2018). https://doi.org/10.15388/infedu.2018.14
8. Chalmers, C.: Robotics and computational thinking in primary school. Int. J. Child-Comput. Interact. **17**, 93–100 (2018)

9. Angeli, C., Xerou, E., Nicolaou, M.: Investigating k-2 students' computational thinking skills during a problem-solving activity about the water cycle using educational robotics. In: 16th International Conference on Cognition and Exploratory Learning in Digital Age (CELDA 2019), pp. 85–91 (2019). https://doi.org/10.33965/celda2019_201911L011
10. Karaahmetoğlu, K., Korkmaz, Ö.: The effect of project-based Arduino educational robot applications on students' computational thinking skills and their perception of Basic Stem skill levels. Particip. Educ. Res. (PER) **6**(2), 1–14 (2019). https://doi.org/10.17275/per.19.8.6.2
11. Luna, A., Chong, M.: A PBL approach for teaching electronics fundamentals by developing robotics projects. In: 2020 XIV Technologies Applied to Electronics Teaching Conference (TAEE), Porto, Portugal, pp. 1–7 (2020). https://doi.org/10.1109/TAEE46915.2020.9163791
12. Piedade, J., Dorotea, N., Pedro, A., Matos, J.F.: On Teaching programming fundamentals and computational thinking with educational robotics: a didactic experience with pre-service teachers. Educ. Sci. **10**(214), 1–15 (2020). https://doi.org/10.3390/educsci10090214
13. Turner, S., Hill, G.: Robots in problem-solving and programming. In: 8th Annual Conference of the Subject Centre for Information and Computer Sciences, pp. 82–85 (2007). https://doi.org/10.13140/2.1.2425.6800
14. Castledine, R., Chalmers, C.: LEGO robotics: an authentic problem-solving tool? Des. Technol. Educ. Int. J. **16**(3), 19–27 (2011). https://ojs.lboro.ac.uk/DATE/article/view/1661
15. Bers, U.: Project InterActions: a multigenerational robotic learning environment. J. Sci. Educ. Technol. **16**, 537–552 (2007). https://doi.org/10.1007/s10956-007-9074-2
16. Scaradozzi, D., Sorbi, L., Pedale, A., Valzano, M., Vergine, C.: Teaching robotics at the primary school: an innovative approach. Procedia Soc. Behav. Sci. **174**, 3838–3846 (2015). https://doi.org/10.1016/j.sbspro.2015.01.1122
17. Fidan, U., Yalçın, Y.: Lego Nxt Training Kit. Afyon Kocatepe Univ. J. Sci. **12**(1), 1–8 (2012)
18. Agatolio, F., Moro, M., Menegatti, E., Pivetti, M.: A critical reflection on the expectations about the impact of educational robotics on problem-solving capability. In: Strand, M., Dillmann, R., Menegatti, E., Ghidoni, S. (eds.) Intelligent Autonomous Systems 15. IAS 2018. Advances in Intelligent Systems and Computing, vol. 867, pp. 877–888. Springer, Cham (2019). https://doi.org/10.1007/978-3-030-01370-7_68
19. Alimisis, D., Alimisi, R., Loukatos, D., Zoulias, E.: Kids make their own robots: good practices from the eCraft2Learn project. Form@re **19**(1), 12–29 (2019). https://doi.org/10.13128/formare-24733
20. Ersoy, H., Madran, R.O., Gülbahar, Y.: A model proposal for teaching programming languages: robot programming. In: Conference of Academic Informatics 2011, pp. 731–736 (2011)
21. Kubilinskiene, S., Zilinskiene, I., Dagiene, V., Sinkevicius, V.: Applying robotics in school education: a systematic review. Baltic J. Modern Comput. **5**(1), 50–69 (2017). https://doi.org/10.22364/bjmc.2016.4.4.01
22. Chalmers, C., Nason, R.: Systems thinking approach to robotics curriculum in schools. In: Kline, M. (ed.) Robotics in STEM Education: Redesigning the Learning Experience, vol. 70. Springer, Dordrecht (2017)
23. Merriam, S.B.: Qualitative Research: A Guide to Design and Implementation, 2nd edn. Jossey-Bass, San Francisco (2009)
24. Garavaglia, A., Petti, L., Murgia, E., Bassi, F., Maranesi, S.: Introduzione della robotica in attività di problem-solving nella scuola primaria. Analisi dei livelli di focalizzazione sugli scopi del problema. Mondo Digitale **75** (2018)
25. Varnado, T.E.: The effects of a technological problem-solving activity on FIRST LEGO league participants' problem-solving style and performance. Ph.D. Dissertation Virginia Polytechnic Institute and State University, Blacksburg, Virginia (2005)

26. Jonassen, D.H.: Learning to Solve Problems: A Handbook. Routledge, New York (2011). https://doi.org/10.4324/9780203847527
27. Berthoz, A.: La scienza della decisione. Codice Edizioni, Torino (2014)
28. Rivoltella, P.C.: La previsione. Neuroscienze, apprendimento, didattica. La Scuola, Brescia (2014)
29. Lucangeli, D., Tressoldi, P.E., Cendron, M.: Test SPM – Abilità di soluzione dei problemi matematici. Erickson, Trento (2003)
30. Laurillard, D.: Teaching as a Design Science. Building Pedagogical Patterns for Learning and Technology. Routledge, London (2012)

Why Educational Robotics May Support Teachers to Discover, to Develop and to Promote Students' Talent: The GIF4T Approach

Lorella Burlin[1], Giordano Casonato[1], Massimo Saccardi[1], and Michele Moro[2(✉)]

[1] GIF – Gruppo Insegnanti Formatori, Oderzo, TV, Italy
{lorella.burlin,giordano,massimo.saccardi}@gifonline.com
[2] Department of Information Engineering, University of Padova, Padova, Italy
michele.moro@unipd.it

Abstract. Talent is present in any person as an innate potential to think, design and realize innovation. Thus it is something worth to be recognized and correctly stimulated. This paper aims to relate the development of talents to educational robotics (ER) as a way to promote creativity, autonomy, computational thinking, project-based and collaborative learning, constructionist approach, gamification. The 'Gruppo Insegnanti Formatori' (Training Teacher Group - GIF) has established the GIF4T initiative to support talent development through ER within the so called T-Labs. After a pedagogical introduction to the talent concept, the paper presents the structure of a T-Lab and its robotic declination. A T-Lab is essentially based on three stages: teaching, experiential learning and talent exploitation through meetings. Then how robots add some relevant enabling factors is explained. The expected results are confirmed by two short case histories in the very specific situation of a remote telematics implementation of the lab (due to pandemic). Some final remarks and ideas of future works complete the contribution.

Keywords: Educational robotics · Talent · T-Lab · Multiple intelligences · Talent profile · Enabling factors

1 Introduction

In the first lines of his recent book [1] Mitchel Resnick anecdotally highlights the current strong need to change education abandoning the 'priority on teaching students to follow instructions and rules' (i.e. to prepare the so called *A students* able to get lots of A grades) and moving to grow students 'willing to take risks and try new things (coming up) with the most innovative ideas and creative new directions' (*X students*). Current and future workplace ask more and more for creativity, adaptability, self- and group-initiative and courage to accept new challenges. In order to make this development from kindergarten to the highest level of education fruitful and particularly felicitous, students should find always attracting and rewarding their activity in and out of school. If we observe how pupils learn in a kindergarten, they show interest in practical activities with situations

M. Malvezzi et al. (Eds.): EDUROBOTICS 2021, SCI 982, pp. 15–25, 2021.
https://doi.org/10.1007/978-3-030-77022-8_2

and objects related to their natural attitudes and often they choose specific roles in which they can express the best of their innate potential.

In our perspective promoting such an innovative approach at any level means to cultivate what we call (personal) *talent*. This keyword should not be confused with other specific personal qualities, worthy to be dealt with special attention, like high potential and giftedness. We are considering one's propensities which can be more or less evident, present in any individual and getting out from his/her innate heritage, social environment and previous education. It is one's own talent(s) that very often act(s) as an engine in early stages and motivate(s) the most positive development both under the practical and cognitive point of view.

This paper aims to relate the development of talents to educational robotics (ER). Literature recognizes the relevance of ER in terms of creativity, autonomy, computational thinking, project-based and collaborative learning, constructionist approach, gamification. We would provide some insights on ER and talent development and promotion, presenting our recent initiative called GIF4T (GIF for talent). GIF (Gruppo Insegnanti Formatori – Training Teacher Group) is a working group of experienced teachers and professionals who organizes training courses and projects for innovating the teaching/learning process. All the preparatory work and case studies developed around ER, talent and distance learning have been condensed in a disseminating publication in Italian [2]. This paper is a meditated summary of its content and it is divided as follows: in Sect. 2 we introduce some pedagogical foundations for our proposal, in Sect. 3 and 4 we present T-Lab and its robotic variation, Sect. 5 contains a short case history, a discussion and future works conclude the contribution.

2 Pedagogical Foundations of Talent Development

The definition of talent which we propose in this paper is inspired by the following scientific studies proposed by various researchers in pedagogy and neuroscience, taking into account the most useful aspects for their applicability to daily teaching, in particular for laboratorial activities. Our specific focus, as previously underlined, is about ER.

- Renzulli [3], in the part of his talent model applied to gifted children, helps us to understand the importance given to creativity in the process of talent development and to motivation as a determining factor for development (Fig. 1).
- Heller [4, 5] emphasizes that there are talent development factors that lead to high performance in different areas of human action. Therefore, talent manifests itself with a performance of excellence. Moreover, we can argue that what we call 'relational climate' favors the development of talent (Fig. 2).
- Gagnè [6] helps us to understand that natural skills are the basis of the development of talent and the development process is achieved only thanks to training and constant commitment to the exercise of natural endowment recognized in formal and informal settings (Fig. 3).
- Gardner [7] clarifies that natural abilities can take the form of different intelligences, citing these nine ones: visual and spatial, kinesthetic, linguistic, interpersonal, intrapersonal, mathematical, musical, naturalistic, existential.

- Margiotta [8, 9] gives us the perspective of the becoming, arguing that talent is the result of a journey; therefore, it is possible to make a forecast, rather than a real identification; in fact, talent is constantly evolving.

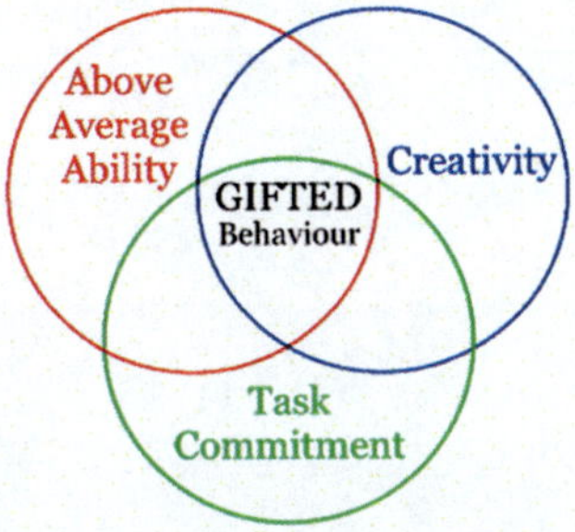

Fig. 1. The three ring conception of giftedness by J. Renzulli [3]

In light of all this, the GIF4T research group has developed a possible definition of talent, to encourage, especially in formal learning contexts such as schools, support for teachers in creating the climate favorable to the blossoming of talents in students.

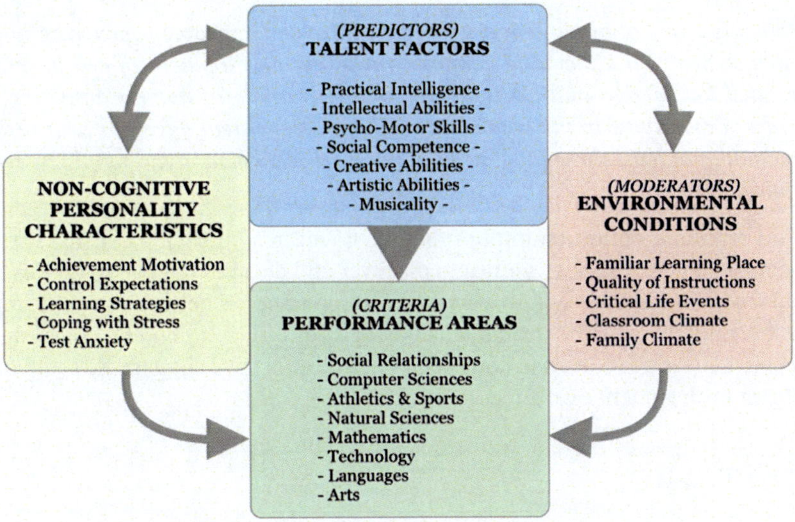

Fig. 2. The Munich model of giftedness by K.A. Heller [4, 5]

As well described in Sir Ken Robinson's TED talk 'How to escape education's death valley' [10], we strongly believe that the role of a teacher is, in increasing form, to be 'climate regulator', not only a provider of knowledge. And the determining factor for creating the right climate is certainly the interpersonal relationship.

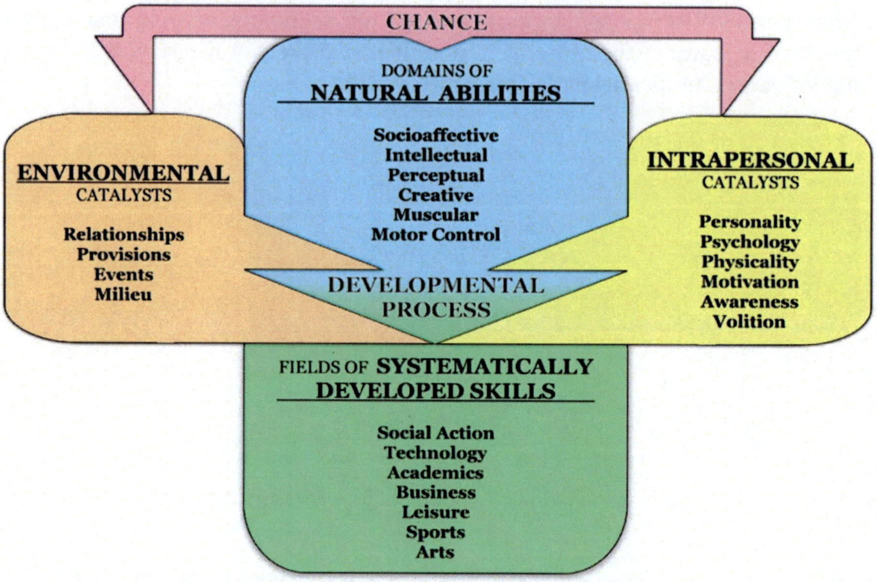

Fig. 3. The differentiated model of giftedness and talent by F. Gagné [6]

From all these assumptions it follows that:

Talent is a mix of intelligence, willing, culture and character, recurring in a person who demonstrates a systematic commitment to the development of competences, with the related knowledge and skills in one or more areas of human activities, through evidence of creativity, in the community of reference. Talent develops as a result of the interaction of at least three factors: multiple intelligence, motivation and interpersonal relationships.

This reference definition implies that the talent in a nutshell exists in every person. The school can create opportunities to discover and develop talents: to do this, teachers should scout any signal, reaction, proposal and spontaneous behavior in each single pupil and coherently drive his/her teaching choosing appropriate individual goals and specific roles [11, 12]. As a fall-out and completion of this effort, we suggest the setup of a *talent profile* for each student over time.

3 The T-Lab

The GIF4T proposal is based on the organization of T-Labs (*talent laboratories*): a T-Lab is an educational tool aimed at facilitating and observing possible evidences of student talent. In fact, individual attitudes are observed in the T-Lab activity in moments in which concepts and theoretical knowledge becomes application in an experiential context. For the student it is an opportunity to know better him/herself by contextualizing his/her attitudes, preferences and aspirations in a community that can recognize them.

The T-Lab is a curricular or extra-curricular practical activity divided into three moments (dimensions): a training phase, an experiential learning phase, moments of contamination with the socio-economic reality of the area. Each intervention involves the creation of a concrete output which can be a product or a service.

How does it differ from a laboratorial experience that is often proposed during the school time? There are three distinctive aspects for the T-lab:

1. The philosophy of talent applied by monitoring its factors;
2. The presence together with the disciplinary teacher of the educator dedicated to talent observation and development;
3. The relationship with talent promoters, experts who we prefer to call *talent masters,* in the socio-economic reality of the territory;

The abovementioned three phases of a T-Lab maybe detailed as follows:

Teaching
Theoretical foundations that support the T-Lab are presented and developed. Though more similar to traditional lectures, anyway this phase should not be instructional, rather conducted with frequent interactions with the students in the perspective of emerging of innate attitudes.
Experiential learning
The practical setup is established, expected outputs explained and agreed, preliminary ideas shared. During this phase, roles are assigned according to the evidences emerged during the teaching phase but the activity progress should be monitored to assure the learning goals and the talent development.
Talent exploitation through meetings
Local talent master(s) meet(s) the students as a moment of relationship-sharing between students and essence of the neighbor productive area in view of an exploitation of talent.

Usually teachers evaluate the student's performance by means of tests and other quantitative assessment tools. In the T-Lab, the result (output) of the student's project work is evaluated by a local "talent master". The T-Lab can, and must be, a bridge connecting the school with the local area, in order to avoid the risk that the school is self-referential.

During the T-Lab activity, teachers have the task of collecting their observations in the student's talent portfolio. The talent portfolio will provide teachers with the information to set up the student's talent profile at the end of each school level. The T-Lab could have the limit of being a single laboratory intervention. We need more than one T-Lab to arrange a student's talent profile. This suggests to plan a series of T-Labs organized in disciplinary areas, which can be connected to different multiple intelligences in order to have a sufficient richness to draw a proven talent profile of the student.

The T-Lab does not replace, but incorporates successful teaching methodologies and techniques oriented to favor evidence of talent, such as a constructionist approach, Inquiry Based Learning, Situated Learning, Flipped Classroom, Cooperative Learning.

4 The Robotic T-Lab

In what extent is it possible to imagine an effective Robotic T-Lab (Robo T-Lab)? Actually ER offers many opportunities to favor the three factors of talent development: Motivation, Multiple Intelligence, Interpersonal Relationship. Motivation is related to the attractiveness of the robotic challenges, simple or complex they might be; multiple intelligence is activated by the action within different disciplines; Interpersonal relationship grows when team-working is stimulated and all aspects of the robotic project (construction, programming, setup of the environment, evaluation) are involved. ER also naturally offers the possibility of working on experiments without an upper limit, thus favoring the expression of creativity, which is one of the most important indicators of evidence of talent.

In deeper detail, five indicators show these enabling factors (Table 1): they have been selected on the basis of our acquired experience in school settings during laboratorial activities.

Table 1. Robotic T-Lab: enabling factors

Factor	Motivation	Multiple Intelligence	Interpersonal relationship
Indicator			
1	Curiosity	Problem solving	Verbal communication
2	Gratification	Innovativeness/Creativity	Teamwork
3	Spontaneity	Ability to process and collect information	Self-control
4	Constancy	Mastery of one's own knowledge	Leadership
5	Interest	Planning	Empathy

Now let's explicit how ER favors observation and development of the three factors of talent.

- *ER and the Motivation Factor*: the building of an artefact (robot), capable of exhibiting human behavior, creates on the one hand an emotional involvement, and on the other one curiosity for other possible actions to be carried out by it.
- *ER and the Multiple Intelligence Factor*: the multidisciplinary nature of ER allows you to field multiple intelligences [7] to achieve the expected results and to reach the final product of the experiment.
- *ER and the Relation Factor*: project- and team-work, typical in a Robotics T-Lab, favor the relationship between classmates and with the teacher-educator.

The talent development educator, who accompanies the laboratory teacher in carrying out the lab, monitors each of the three talent factors through the observation of these indicators.

In Sect. 3 we highlighted the necessity to organize a series of T-Labs, covering different disciplinary areas, in order to promote multiple intelligences and to arrange a meaningful student's talent profile. Though a Robotics T-Lab could be just one of this series, we would like to prove that a Robo T-Lab offers in itself the advantage of designing activities that are intrinsically multidisciplinary, therefore allows teachers to investigate all multiple intelligences.

5 Two Short Case Histories

In March 2020, the COVID-19 emergency forces everyone to lockdown with consequent social distancing. Everyday life undergoes a sharp change. Also in Italy work and school had to rethink their organization quickly and not painlessly [13]. A comprehensive institute in the Treviso district wanted to put itself into play to support young people in this difficult moment. The school agreed to experiment a sort of Remote RE that we called with the acronym *REaD* (*Robotica Educativa a Distanza* – Distance ER). The lab was proposed by the GIF trainers to a group of 25 students of the junior secondary school, from 10 to 12 years old. The teacher experienced in educational robotics was one of the GIF researchers involved in the GIF4T-approach project. The tutor-educator was one of the teachers of the school, who had the task of observing the behavior of the young students.

As a preliminary proof of the effectiveness of our proposal, we present two case histories showing how a Robo T-Lab can be organized and which result you can expect in terms of talent development.

During this experiment, as well as during other previous ones, we could observe evidence of students' talent which had not been observed in traditional lectures or other kinds of laboratory activities. Moreover, we choose these two particular cases, mainly because they tell us about two different multiple intelligences: logical-mathematical and interpersonal. This is a further proof of how educational robotics may investigate different personal attitudes in students working in teams.

5.1 The Creativity of Maria

Talent development factor: multiple intelligence
Multiple intelligence indicator: problem solving

Challenge

This experience used the Open Roberta Lab Virtual system [14] that includes a block-based programming interface and a simulator with a some virtual environment where to move with obstacles, colored areas and markers. The ER teacher, after providing the necessary instructions for programming the simulated autonomous robot, decides to launch an ambitious robotic challenge for her pupils. The robot must reach a green lawn, identifying the correct path. The virtual setting features black tracks, with sometimes sharp bends and crossroads marked with green and red squares. Time allowed: half an hour.

The challenge is considerable, the teacher has given a task that requires more skills and abilities than those provided in the initial phase of familiarization with the virtual robot and the programming language.

Solution

When time runs out, she asks the kids to share the problems they encountered in facing the robotic challenge. Therefore, the teacher does not expect the solution, on the contrary, she fears that she has overestimated the skills of his young students and therefore asks the children to explain the problems encountered.

Maria, usually silent working with the camera switched off, asks in the class chat to be able to share her solution. Maria solved the problem brilliantly! Her solution was not only based on writing an excellent program for the robot, but also on a constructive strategy. She equipped the robot with three front light sensors instead of just one. Maria explains that with the three sensors the robot is able to distinguish the green lawn. She claims that if the three sensors "see" the green color, the robot will know it has reached the lawn. If the robot were instead at the crossroads, only one sensor would see the green color.

The teacher is speechless for a few seconds. An unexpected success!

The classmates begin to ask for explanations directly from Maria, who activates her camera and takes the teacher's place by responding directly to the classmates.

Benefits

Maria, after this experience, becomes aware of one's problem solving skills. She begins to work with greater self-esteem, cultivating her inclination. Maria also begins to understand that it is important to share one's skills with her fellows and finds gratification in helping others.

Educational Reflection

Objectives achieved by Maria:

1. Didactic: The robot reaches the green lawn therefore Maria acquires the knowledge and competence required by the difficulty posed by the task.
2. Talent: Maria begins to understand her value in terms of logical-mathematical intelligence, she feels on the path of improving her relational skills and reaches a higher degree of motivation in the themes introduced by the REaD Laboratory.
3. Educational: Maria begins to get involved with her classmates after solving the educational challenge posed by the teacher and therefore rediscovers self-esteem and new awareness of herself and of the personal resources she possesses and that others recognize them.

The teacher has therefore succeeded with Maria in the intent launched with the educational challenge: the Remote Robo T-Lab the facilitation and observation of evidence of talent will allow the teacher to contribute to the definition of the student's talent profile.

Conclusion

The REaD experimental laboratory made it possible to discover Maria's talent in a nutshell, thanks to her creative found solution. We can therefore conclude that precisely because of this trait of innovativeness, we observe the evidence of a logical-mathematical talent that goes beyond the correct execution of the task, that is, a best performance of computational thinking.

5.2 The Empathy of Marco

Talent development factor: relationship
Relationship indicator: empathy

Within a ER project funded by PON[1], we tell the story of Marco, one of the participating students attending a comprehensive school. The boy entered the first grade of the junior secondary school with ADHD (Attention Deficit Hyperactivity Disorder) certification. This disorder had led him, especially in the primary school years, to live the relationship with his peers and with adults with difficulty due to his impetuous behavior.

Despite this he decides to join the lab, even in its new remote version. From the very first moment he is motivated and enthusiastic, made evident by his constant presence, even though it is an activity that forces him to sit on a chair, in front of the computer, all along two full hours.

Challenge

A somewhat demanding challenge for the education team! The expert teacher and the observing tutor must face the risk that Marco may give up. The distance learning may present some problems that can discourage or make the didactic activities a bit difficult: poor connection, sudden interruptions, difficulties in accessing meetings, slowness of home (and, sometime, even school) computers just to mention some, put a strain on the patience of even the most motivated.

Solution

With everyone's surprise, Marco constantly offers his help to fellow students in difficulty through the chat system in order not to interrupt the lesson in his turn. The boy, who declares himself a lover of computer science in general, through his sometimes funny and theatrical ways, never fails to offer support or help in solving technical problems. He gives his classmates suggestions on using the platforms, takes care to contact them personally when he does not see them present in the class and talks to the teachers about any difficulties that prevent the participation of some classmates. On more than one occasion, he also supports their point of view with the teacher and repeats the explanations to them.

Marco thus demonstrates that he is suitable for interpersonal relationships, which becomes evident from the empathy shown in helping classmates.

Benefits

Marco, after the T-Lab experience, sees himself with new eyes. He revitalized his relationship with his peers, often becoming a point of reference. Moreover, they have learned to like his way of being and to smile at his remarked expressiveness, due to his impetuous behavior. Now Marco is becoming more and more aware about his emerging interpersonal talent and begins to work for himself and for others with greater self-esteem and ease. Furthermore, he starts to embark on a journey of greater awareness of his skills and competences, that are both logical-mathematical and interpersonal.

---

[1] PON – *Programmi Operativi Nazionali* – National Operative Programs, a ministerial initiative which uses European funds for development to support renovation of the school system.

6 Discussion and Future Works

The lesson learnt around the presented case history is this: remote ER, as well as in the presence, allowed the observation and development of a deep educational process in Maria and Marco. In our opinion this happened:

1. thanks to the characteristics of the robotics laboratory that offered the possibility of working on experiments without an upper limit, thus favoring the expression of creativity and empathy, which are two of the most important indicators of the talent factors;
2. thanks to the active presence of the educator alongside the robotics teacher who was able to collect the evidence of talent shown by the two students.

The system of indicators helped the educator in the process of observation and interaction. The collected evidences can be reported in a useful talent profile: the starting point of a journey towards the full development of talent. Moreover, the positive, we would say essential, contribution of the educator makes it clear the necessity of defining the role and the professional preparation of this new figure that we would like to define as "talent educator" who collaborates with the teacher in the Robo T-Lab.

Educational robotics has been confirmed as a strongly effective educational methodology [15] for the discovery and development of logical-mathematical intelligence and has also offered the possibility of collecting evidence of interpersonal intelligence.

The GIF4T research team has adopted the practice of 'case writing' in educational and pedagogical work, focusing on its use by education professionals as a research and self-reflective strategy. This practice requires a research stance and a reflective and critical outlook by educational professionals [16]. Further relevant case histories are presented in the semi-annual publication named "Talent Diaries" that are collecting successful teachers' experiences in T-Lab.

The difference between a traditional ER laboratory and a Robo T-Lab is in the way of monitoring the result. In the first case the teachers verify the computational thinking level and the success of the robotics experiment. In the second case the teacher and the educator work in a team to investigate the performance requested by the task assigned and also the attitudes of each student in the different intelligences, to collect evidence for the talent profile. A member of the GIF4T team, experienced in both kinds of laboratory activities, may confirm the educational value of T-Lab in order to collect more useful information about the students' attitudes.

The GIF4T-approach is being applied in various kindergarten, primary, junior and senior secondary schools in Italy. The GIF4T researchers argue that the approach can be adapted to all the other school systems, as well as they did successfully with an high school based on the International Baccalaureate (IB) system, the GSO International School of Talent (Veneto-Italy).

The GIF4T working group aims to continue the experimentation launched at the beginning of 2020 to validate the indicators and tools for observing talent factors in schools of all levels. It also intends to deepen the educational dynamics within the three phases of the T-Lab to better understand the educator-teacher-student relationship. We are strongly engaged in the preparation of the structure of the personal talent profile, convinced that collecting these evidences with practical experimental activities in school could be beneficial for what will happen after leaving the school system.

References

1. Resnick, M., Robinson, K.: Lifelong Kindergarten: Cultivating Creativity Through Projects, Passion, Peers, and Play. MIT Press, Cambridge (2017)
2. Burlin, L., Casonato, G., Saccardi, M.: Laboratorio di talento con la robotica educativa a distanza, Quaderni del Talento (The Talent Lab Based on Educational Robotics, Talents Diaries), vol. 1, La Musa Talìa, Venice, Italy (2020). (in Italian)
3. Renzulli, J.S.: Schools for Talent Development: A Practical Plan for Total School Improvement. Prufrock Press (1994)
4. Heller, K.A., et al. (eds.): International Handbook of Giftedness and Talent. Elsevier, Amsterdam (2000)
5. Heller, K.A., Perleth, C., Lim, T.K.: The Munich model of giftedness designed to identify and promote gifted students. In: Conceptions of Giftedness, vol. 2, pp. 147–170 (2005)
6. Gagné, F.: Talent development as seen through the differentiated model of giftedness and talent. In: Balchin, T., Hymer, B., Matthews, D.J. (eds.) The Routledge International Companion to Gifted Education, pp. 32–41. Routledge (2013)
7. Gardner, H.: Formae mentis. Saggio sulla pluralità dell'intelligenza. Feltrinelli, Milano (1987). (in Italian)
8. Margiotta, U.: La formazione dei talenti: tutti i bambini sono un dono, il talento non è un dono (Talent Training: All Children are a Gift, Talent is Not a Gift). F. Angeli, Milano (2018)
9. Morselli, D., Costa, M., Margiotta, U.: Entrepreneurship education based on the Change Laboratory. Int. J. Manag. Educ. **12**(3), 333–348 (2014)
10. Robinson, K.: How to escape education's death valley. TED talk (2013). https://www.ted.com/talks/sir_ken_robinson_how_to_escape_education_s_death_valley/transcript
11. Feldman, D.H.: Beyond Universals in Cognitive Development. Ablex Publishing, Norwood (1994)
12. Pivetti, M., Di Battista, S., Agatolio, F., Simaku, B., Moro, M., Menegatti, E.: Educational robotics for children with neurodevelopmental disorders: a systematic review. Heliyon **6**(10), e05160 (2020)
13. Resnick, M.: Cultivating Creativity During the Coronavirus Crisis, MIT Media Lab https://www.media.mit.edu/posts/cultivating-creativity-during-the-coronavirus-crisis/. Accessed 17 Nov 2020
14. Fraunhofer Inst.: Open Roberta Lab. https://www.roberta-home.de/en/lab/. Accessed 17 Nov 2020
15. Moro, M., Agatolio, F., Menegatti, E.: The RoboESL project: development, evaluation and outcomes regarding the proposed robotic enhanced curricula. Int. J. Smart Educ. Urban Soc. (IJSEUS) **9**(1), 48–60 (2018)
16. Biffi, E.: 'Case writing' as a research strategy for the educational professions. Encyclopaideia XVIII (39), 117–134 (2014). ISSN. 1825-8670

Robotics Education Under COVID-19 Conditions with Educational Modular Robots

Rico Möckel(✉), Lucas Dahl, Dean Boonen, Desirée Parren, and Yiyong Gou

DKE SwarmLab, Department of Data Science and Knowledge Engineering, Maastricht University, Maastricht, The Netherlands
rico.mockel@maastrichtuniversity.nl

Abstract. The COVID-19 pandemic forces many robotics teachers to rethink their approach to education. Distance rules and the constant threat of a partial or complete lockdown leading to limited access to classroom equipment make it challenging to plan for hands-on education where students experience robotics by experimenting and studying with robotic hardware. On the other hand, this hands-on active learning experience is one of the strengths of robotics education and the ability to handle hardware equipment a substantial learning goal of study programs on robotics. In this paper we present and discuss the approach taken for the course Robotics and Embedded Systems at Maastricht University. The course had been adjusted to meet COVID-19 safety regulations and to allow for a fast seamless transition between onsite education at university and online education where students can work with robotic hardware at home. We share experience, best practice advice as well as educational material to help other teachers benefit from our developments. A key contribution is our custom-made, low-cost, educational modular robotic system for teaching kinematics, locomotion, and PID control that we make publicly available for replication through the website https://www.maastrichtuniversity.nl/edmo.

1 Introduction

In robotics education, students benefit from hands-on experience with robotic hardware that helps students engage in an active learning process [1] where students apply, test, and gain new knowledge by experimenting on real-world systems. This hands-on experience is beneficial for their learning and understanding as students learn best when they (I) experience course content in multiple modalities, (II) connect new knowledge to existing knowledge, (III) focus their attention on the learning experience, (IV) stay motivated to learn, (V) receive appropriate feedback early and often, and (VI) work on engaging collaborative and authentic experiences [2–4] - modalities that are facilitated by lab sessions where students actively interact with robotic hardware.

R. Möckel and L. Dahl—Both authors contributed equally.

M. Malvezzi et al. (Eds.): EDUROBOTICS 2021, SCI 982, pp. 26–38, 2021.
https://doi.org/10.1007/978-3-030-77022-8_3

However, due to the COVID-19 pandemic and the consequentially required safety measures to protect the health of our students and teachers this core concept of robotics education, the intense hands-on experience, faces serious challenges. These challenges range from social distancing (i.e. keeping a minimum distance of 1.5 m) during all educational activities to not being able to disinfect electronics and several types of robotic materials making it hardly possible for students to share robotic equipment and thus debilitate group work. Also due to the constant threat that countries or at least some regions have to undergo a complete or partial lock-down at short notice, educational institutions and teachers must be prepared to quickly adapt education to scenarios where onsite education in class is no longer feasible and students have to be taught at home. As a consequence of all these conditions, robotics education must be designed in such a way that a quick adaptation is possible while maintaining high educational standards and while meeting intended learning outcomes.

This paper describes the approach of and experience from the course "Robotics and Embedded Systems" of the Bachelor study program "Data Science and Artificial Intelligence" at Maastricht University, The Netherlands. The course has been taught during the COVID-19 pandemic from beginning of September until the end of October 2020. The course had been heavily adjusted to successfully address the aforementioned challenges of robotics education under COVID-19 conditions. With this paper, we share our experience in teaching robotics in higher education and provide other teachers with advice and open-source educational robotic hardware. We share experience and hardware in the hope that this provides meaningful insights and support for other educational institutions and colleagues. In particular, we are describing how we adapted the course to meet the conditions of the COVID-19 situation and to be able to switch between onsite and online education at short notice. A key tool that allowed us to maintain high safety standards and to let all our students study with their own robotic hardware is our custom-made, low-cost, robust modular robotic system EDMO. EDMO stands for EDucational MOdular robot as we developed the EDMO system particularly for robotics education. EDMO can be rapidly replicated through 3D printing and by combining low-cost off-the-shelf components. These are ideal properties if, like for our course, a large number of robotic setups have to be created to provide all students with personal hardware equipment to meet COVID-19 safety regulations. With this paper we extend our previously published EDMO setups [5] by additional and upgraded modules that we again make openly available for replication through the website https://www.maastrichtuniversity.nl/edmo.

2 Course Overview and Intended Learning Outcomes

An overview of the main components of the course on "Robotics and Embedded Systems" is provided in Fig. 1. The core of the course is formed by tutorials and lab assignments that allow students to practice with electronic and robotic hardware. We let students practice with hardware, rather than simulations, so that

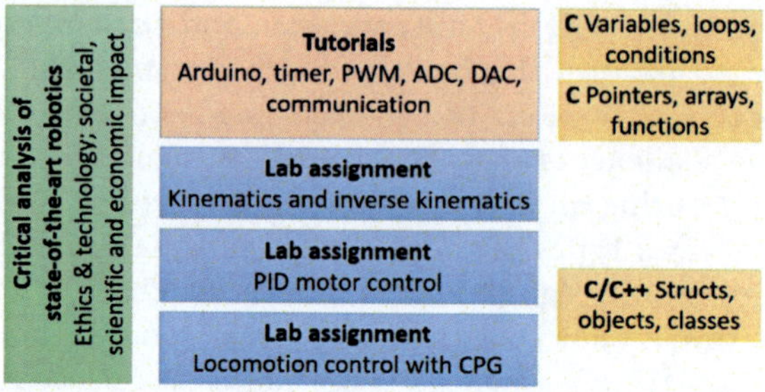

Fig. 1. Overview of course elements and content.

students learn to create robust robotic control systems and to deal with physical effects like friction, backlash, and noise, which is one of the important intended learning outcomes (ILOs) of the course. Further ILOs - addressed by tutorials, labs, and lectures introducing knowledge on robotics, control, and mechatronics - include the ability to program embedded systems under real-time constraints for reading sensors, communication, and robot control. After successful conclusion of the course, students are able to apply three main control concepts: (I) control of kinematics and inverse kinematics [6] of a robot arm, (II) closed-loop PID motor control [7], and (III) locomotion control with central pattern generators [8]. Thus students can apply a variety of useful, widely used control concepts. Students also gain a better understanding of important techniques from mathematics and artificial intelligence including parameter identification and optimization as well as the application of (coupled) differential equations and numerical mathematics for modeling and controlling robotic systems. Next to control concepts, students receive a step-by-step introduction to learn to program in C/C++, an ability that students directly practice as part of tutorials and lab assignments. Every week students get introduced to additional state-of-the-art robotic systems that are critically analyzed and discussed regarding their technology, ethical implications as well as their societal, scientific, and economic impact. We included these discussions into the course because we want our students to be well aware of the impact of robotics and artificial intelligence on our society so that students can make ethically well-justified decisions during their professional career.

Despite the impact of COVID-19 conditions we were determined to maintain meeting all ILOs of the course. That included also those ILOs described before that require hands-on experience with electronic and robotic hardware. To meet all ILOs of the course and to prepare properly for a possible lock-down that would only allow for online teaching, we upgraded both our EDMO hardware and our teaching approach. Both adjustments are being discussed in the following sections.

3 EDMO Hardware Setups

During the course, each student was provided with three custom-made EDucational MOdular (EDMO) robot modules (EDMO-RM, Fig. 2b) and one custom-made EDMO DC motor module (EDMO-MM, Fig. 2g). Together, EDMO-RM and EDMO-MM form the educational modular robotic platform EDMO. All EDMO modules have been designed to meet the requirements of low costs production with small size, low-cost hobby 3D printers. The printing process requires only a single extruder and no support material.

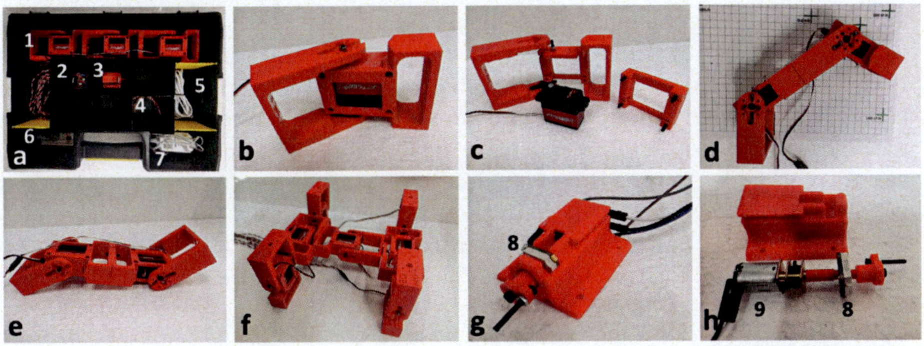

Fig. 2. EDMO hardware: (a) EDMO box ready for shipping to students. Box contains {1} three EDMO-RM, {2} DC motor driver, {3} EDMO-MM, {4, 5} cables, {6} microprocessor and breadboard with electronic components, and {7} a DC power supply. (b) Assembled and (c) disassembled EDMO-RM. (d) EDMO robot arm made of EDMO-RM for studying control of (inverse) kinematics. (e) Snake configuration for studying locomotion control. (f) EDMO quadruped. (g) Assembled and (h) disassembled EDMO-MM containing {8} a rotary encoder and {9} a DC motor with gearbox.eps

3.1 EDMO Robot Modules

EDMO-RMs (Fig. 2b) were designed to create robust, versatile robotic modules from which a variety of robotic morphologies can be achieved including robot arms (Fig. 2d) for studying control of (inverse) kinematics and snake-like structures (Fig. 2e), but also full biped and quadruped configurations (Fig. 2f) for studying locomotion control. Also due to the modular design approach, without in-depth knowledge on the mechanic design of a robot, students can easily repair non-functioning robot configurations by exchanging malfunctioning robot modules.

Within a larger configuration, each EDMO-RM forms one module with a single hinge joint. Figure 2c shows a disassembled EDMO-RM. EDMO-RMs combine a low-cost servo motor integrated into a custom-made 3D printed body and were designed so that their assembly and the replacement of the servo motor requires little skill. As a result, educational institutions can i.e. perform the production of EDMO modules at low costs with the help of student assistants that do not need to have a technical background. Alternatively, the assembly of EDMO modules can become an integrated part of a robotics course where students get asked to assemble those EDMO modules, that they are afterwards using during their lab assignments. We updated EDMO-RM version 1.0 (originally published in [5]) and created version 1.1 around the TrackStar TS-411MG digital servo motor with metal gearbox. Featuring a torque of 11.2 kg·cm, the EDMO-RM version 1.1. is five times stronger than EDMO-RM 1.0 at an additional cost of €10.10. Due to the extra torque, EDMOs now support the construction of larger robot configurations and longer robot arms. Both EDMO-RM versions can be operated together in a single robot configuration. During the course, both EDMO-RM versions proofed to be robust in their application. All experiments could be carried out with only two EDMO-RM. We provided all students with three modules to allow them to quickly repair a robot arm or snake in case of a module failure. However, despite intensive use by about 40 students, no EDMO-RM was damaged. So in the end, students used replacement modules only to try larger and more sophisticated robot configurations, not to replace damaged modules.

3.2 EDMO DC Motor Module

In past years we have tested a variety of hardware configurations for education on PID motor control. A key challenge during implementation of PID control is that students have to take care of mechanical limits of servo motor components. Our strategy in past years involved supplying our students with low-cost servo motors from which we had removed the motor control electronics. Students would then read the angular position from the potentiometer inside the servo motor and drive the embedded DC motor through a separate motor driver board (L298N). If students would break some internal components of the servo motor, they would learn from this experience and would have to start with a new servo motor. With the risk of running into a shutdown of the university, we found this strategy of supplying replacement motors too risky and went for designing a new custom-made 3D-printed DC motor module, the EDMO-MM. EDMO-MM is designed to be very robust by combining a DC motor with metal gears with a N15TS-103A3030 rotary encoder from Amphenol Piher (Fig. 2g, h). By design, EDMO-MM supports continuous rotation and is operating without those mechanical limits that are typically found in standard position-controlled servo motors and that led to servo motor damages in past teaching. As a result, with EDMO-MM we did no longer encounter any motor failure during the course, despite intensive use, just two cases of not properly soldered cables that could be quickly fixed and can be avoided in the future.

3.3 EDMO Box

Situated in the heart of Europe and within 30 km from the German and Belgian borders, Maastricht University attracts students that commute across country borders on a daily basis. The result is an international educational environment that is highly appreciated by both students and staff. However, during the COVID-19 pandemic where European countries prefer their citizens to avoid cross-border travels and where we experienced that safety and travel regulations can spontaneously change during a weekend, this forced us to develop additional strategies to ensure that all students would remain access to robotic equipment in case of strict travel restrictions. Also, several students from within the Netherlands faced the problem that they spontaneously could no longer come to class as they had been instructed to quarantine by Dutch health-care authorities at short notice.

Even if onsite education in class could have been maintained, during our course preparations it was not clear how long the COVID-19 virus can survive on surfaces of robotic and electronic hardware. Standard disinfection strategies are not applicable to electronics and robotic hardware since e.g. typical disinfection fluids and sprays can harm electronics and dissolve 3D-printed material. As a result, after careful consideration of alternative strategies we decided that our best option was to provide all students with their own hardware setups. Here we benefited from the chosen robotic and electronic setups that support rapid fabrication at reasonable costs. Through this rapid fabrication we have been able to provide all 41 students of our course with personal robotic setups. No sharing of hardware between students was necessary. Hardware production time was less than one month with a team of only three people. An item and cost list highlighting the main components as well as an upper cost limit can be found in Table 1. A detailed bill of materials is provided online at https://www.maastrichtuniversity.nl/edmo. We found that we can provide each student with a full robotic setup at a maximum costs of 145€ per student. Total costs can be further reduced i.e. by using mass production or by purchasing off-the-shelf components in large volume.

The small form factor of all hardware components allowed us to integrate and properly organize them into off-the-shelf plastic boxes (Fig. 2a) with a size of 44.2 cm × 33.2 cm × 9.2 cm as they are offered by standard hardware shops for storing and organizing hardware items. The box was chosen so that it could be shipped by mail to students that are prohibited to come to class. During onsite education, the outside surfaces of the box could be disinfected by students and staff while only students would have access to material inside. Thus, this organization would allow meeting safety standards during onsite education and would have enabled us to rapidly switch to online education without dropping intended learning outcomes that require hardware experiments.

Table 1. Material costs (upper limits) for EDMO box and components

Component	Costs [€]
EDMO 1.0 robot module	
3D-printed body parts and fasteners	4.50
Turnigy^{TM} TGY-50090M servo motor	5.50
EDMO 1.1 robot module	
3D-printed body parts and fasteners	4.50
TrackStar^{TM} TS-411MG servo motor	15.60
EDMO DC motor module	
3D-printed body parts	3.00
RS PRO 6 V brushed DC geared motor	15.20
Amphenol Piher N15TS-103A3030 rotary encoder	2.40
Electronics	
Adafruit Feather M0 Proto board	18.00
Adafruit Feather servo wing	9.00
L298N DC motor driver board	5.00
5 V DC - 2 A Raspberry Pi power converter	6.80
Breadboard, resistors, LEDs, switch, cables, and wires	16.80
Plastic box/organizer	20.00
Screwdriver	4.00
Total costs of box with 2× EDMO 1.0 robot modules	**<125**
Total costs of box with 2× EDMO 1.1 robot modules	**<145**

4 Adapted Teaching Approach to Meet COVID-19 Conditions

Until before the COVID-19 pandemic, the course featured interactive onsite lectures, tutorials, and lab sessions. During tutorials and lab sessions teaching staff was moving in between students and joined them in debugging code and experimenting with robotic hardware. To challenge interested students and to provide additional support for those students requiring extra help, we provided a personalized learning environment. Personalized learning involves providing all students with a range of different opportunities for understanding new information and developing teaching materials so that all students of a course can learn effectively, regardless of differences in their ability. We applied personalized learning e.g., by inviting all students to our robot lab, the DKE SwarmLab, where students struggling with tutorials or lab assignments got extra practice time and guidance. Excellent students got opportunities to explore and study material that went beyond core content of the course. Often students would also implement own project ideas with support by teaching staff.

When preparing the course during the pandemic we knew that we would no longer be able to maintain this open teaching approach where students could come to the lab whenever they wanted while properly considering COVID-19 safety restrictions. To provide all students with the opportunity to gain some onsite study experience at university including personal contact to teachers and fellow students, Maastricht University followed the guideline: "On campus, if possible, online, if necessary." To prevent infections, students had to keep a minimum distance of 1.5 m at all times considerably reducing the maximum amount of students allowed in university buildings and lecture rooms. To meet these safety guidelines, classes had been split into smaller groups and students were no longer allowed within campus buildings outside scheduled class hours. Thus, concepts like our open lab were no longer possible. In addition during the pandemic, all courses must plan for the constant threat of a transition to fully online education that might come at short notice. During the first wave of the COVID-19 outbreak in Europe before summer 2020, we experienced that during a lockdown students travel to their home countries to escape loneliness. Since our study program proudly attracts international students from all continents we had to prepare education for students suddenly living in different time zones.

When replanning our course we asked ourselves how we can best get students engaged in active learning by experimenting with robotic hardware while closely following COVID-19 safety guidelines as well as prepare for a rapid switch between onsite to full online education if that would become necessary. Also, we wanted to use the limited contact hours at class (and online) in such a way that students benefit most from guidance by experienced staff. Finally, we wanted to maintain an effective form of personalized learning that targets the different needs and strengths of our students. To achieve these goals we reorganized the course following a variation of blended learning (BL) [9] and the flipped classroom approach [10]. BL is known for aiming to optimally integrate face-to-face teaching with online learning [9] with the goal of "using the web for what it does best and using class time for what it does best" [11]. BL does so by replacing or enhancing aspects of face-to-face classroom learning by online or technology-based experiences [9].

In our course, we combined elements of online learning with onsite hands-on tutorials and lab sessions supported by online group discussions and personal onsite discussions. For interactive online communication we used video conference tools like Zoom. Also, all onsite tutorial, lab, and discussion sessions had been planned in such a way that teaching staff could support students online through video communication while students would be working at home if a transition to full online education would become necessary. Course content was provided for self-studies online through a CANVAS platform in form of video lectures and reading material. The course was carefully planned so that students had enough time for self-studies and labs.

To optimize required contact time between teaching staff and students and to limit the number of required teaching assistants during tutorials and lab assignments, also for tutorials and labs we provided short videos with step-by-step instructions as well as videos with background information that students could watch on demand. For each tutorial session, basic and challenging tasks were provided targeting the different skill levels of students while ensuring that all students could reach the intended learning outcomes of the course. This organization of tasks in addition to providing all students with own hardware setups and by having teaching staff available to answer individual student questions led us to create a personalized learning environment [12] for all students that could be seamlessly transferred to full online education. Due to our experience from past years, we had been able to foresee and address many student questions in our video lectures. As a result, online and onsite contact hours could be focused on helping students who needed extra support or had additional questions.

A standard tutorial session was organized as follows: Each tutorial was separated into smaller sub-tasks and experiments. In preparation for a tutorial, students watched short introductory lecture videos at home and studied the required background material. Lecture videos were organized to address students with different background knowledge. All students were asked to study the core material. Additional background material was provided in text and video form to support the understanding of the core material where we expected that some but not all students would require additional information. Also during tutorials, students were watching short videos containing instructions using headsets to not disturb colleagues. There was a separate video for each sub-task. Each tutorial started with short, simple tasks that would become increasingly complex to provide an easy but rewarding start as well as some complex material for more experienced students. For the teaching staff, it initially felt strange to see students watch video instructions silently and to conduct experiments on their own without the involvement of any staff. However, we quickly found that this approach was rewarding both for students and staff. Students followed the tutorials at their own pace without feeling the pressure to keep up with or to wait for colleagues as they had during tutorials before COVID-19. Staff benefited from close interaction with students on interesting questions that often went beyond the standard lecture material, rather than answering basic questions repetitively. Students asked freely since they did not feel observed by colleagues. Also because students went at their own pace, students did not finish a sub-task at the same time. Thus, bottleneck situations where all students required support at the same moment could be reduced, a situation that we encountered more frequently before using video instructions in tutorials and labs. This unloads teachers and enables better communication with students.

During course preparation, we feared that video lectures used by students for self-studies according to the flipped classroom approach [10] would be less motivating for students to follow than the interactive lectures that we used to provide before COVID-19 and that as a result, students might delay their studies to a point where an effective exam preparation would no longer be possible. However, we found that hands-on tutorials and lab assignments significantly helped and motivated students to stay on track since to successfully work with electronics and robotic hardware, students had to study a well-defined amount of video lectures every week. Providing both basic and challenging tasks for all lab sessions helped significantly both students and staff to (self-)assess individual study progress and to identify content that required additional explanations. Typically students could remove study delays by themselves by concentrating on the essential parts of tutorials, skipping advanced content, and by gaining extra support by teaching staff. Since all students worked with their own hardware setups supported by video tutorials, we could provide enough flexibility so that students could shift content from one week to the next where required. In a few cases where students could not catch up and did not meet study progress requirements (typically caused by external circumstances, partially due to private COVID-19 situations) we offered an additional lab session targeted more specifically on the needs of the individual students and with intensified support by teaching staff.

Individualized hardware provided to each student played an important role in fraud prevention. Due to individualization of hardware setups, e.g. by using different motor gear ratios, different alignments of servo motors and rotary encoders, control parameters could not be copied one on one from one hardware setup to another. Students thus had to work intensively on their setups to be successful.

During direct contact, students and staff have been wearing face masks and kept a distance. We also experimented with the use of video lecture tools like Blackboard Collaborate Ultra and Zoom to allow staff to ensure the correct wiring of electronics from a distance and to allow students sharing their computer screen either with teaching staff or to help each other with coding problems. In case of a lockdown where students would no longer have been able to come to class, we would have used these tools to keep supporting students at home.

Students were graded based on two lab assignments (PID control and locomotion control) that counted in total for 20% of the final course grade as well as a written paper exam counting for the remaining 80%. Depending on their preferences and conditions, students could take the final written exam onsite or as a proctored online exam. Lab assignments were graded based on software code and experimental results that students submitted in form of a video. We decided for video submissions as we made good experience with them in the past years and as they (I) allow good protection against fraud as students are asked to comment and explain their experiments with their voice and (II) allow for the documentation of moving robots. (III) In addition we chose for video submissions as they become a more and more common addition of scientific publications i.e. in robotics journals and conferences. Thus, it is essential that also students gain the skill to prepare such video submissions (course ILO).

5 Discussion and Conclusions

In this paper, we share our experience in teaching robotics and embedded systems under COVID-19 safety conditions. We adapted to a flipped-classroom approach with elements of blended learning and used compact, robust, custom-made educational robotic modules, called EDMOs. By this, even under challenging COVID-19 safety regulations, students achieved those intended learning outcomes that require experimenting with robotic hardware. The paper presents the design and use of two new EDMO modules: a second-generation EDMO robot module with hinge joint and a new custom-made EDMO DC motor module for rotatory movements. Through their modular design and by supporting a variety of configurations, EDMO modules allow students to explore, study, and practice approaches for position control, control of (inverse) kinematics and locomotion. We further use EDMOs to teach the application of mathematics and artificial intelligence. All hardware discussed in this paper is made publicly available for reproduction at other educational institutions. The educational approach and material described in this paper are specifically designed to allow for a seamless switch between onsite and online education. Video lectures would allow students to follow the course even in different time zones while all robotic and electronic equipment can be packed so that it could be shipped to individual student homes in case of a total lockdown where teaching at university buildings would no longer be possible.

While teaching robotics with EDMO setups during the COVID-19 pandemic, we found that students embrace the opportunity to have their own hardware setups and to use these setups also for independent problem solving. Providing individual setups for all students was possible for us due to our custom-made EDMO modules that we could reproduce rapidly with few staff members at reasonable costs. Although initially skeptical, we were positively surprised by how well education with robotic hardware is possible despite strict safety regulations. However, preparation time for the adaptation of a robotics course to the new format and for providing sufficient educational material is not to be underestimated. Throughout the course, our students acted in a very professional and responsible way. Students were very understanding that the teaching format had to be adapted. Students also agreed to wear face masks to protect each other and teaching staff even before strict regulations had been passed. Sometimes we had to remind students to keep a distance. However, this is very natural during communication with friends and colleagues.

After taking the course, students at Maastricht University are invited to anonymously evaluate the course, its components, and teacher performance on a 5-point Likert Scale [13], where a grade of 5 corresponds to the highest level of student satisfaction and is rarely given by students. The results of the student evaluations for the presented course are provided in Table 2. Comparing student satisfaction during COVID-19 (course grade of 4.7) and before COVID-19 (average course grade of 4.42) shows the success of the course adaptation. Student satisfaction increased in comparison to last year and slightly outperformed the best-reached value so far from the year 2017. Also feedback by students in personal discussions was very positive. No intended learning outcomes had to be dropped and the level of difficulty of examinations remained at the same high standards we achieved before COVID-19. Average performance of students dur-

Table 2. Course evaluations provided by students.

Academic year	# students who took the course	# students who evaluated the course	Course grade [Likert scale]	Standard deviation (course grade)	Teacher grade [Likert scale]	Standard deviation (teacher grade)
2015	28	16	4.5	0.63	4.88	0.34
2016	23	4	4.33	0.47	4.75	0.43
2017	41	22	4.68	0.55	4.91	0.28
2018	24	9	4.4	0.9	4.6	1.0
2019	47	28	4.2	0.8	4.1	1.0
Mean evaluation - before COVID-19						
	32.6	15.8	4.42	0.67	4.65	0.61
Evaluation 2020 - during COVID-19						
2020	41	21	4.7	0.5	4.8	0.4

ing the final written exam was similar to previous years. However, the passing rate of students taking the exam decreased from 91.7% (before COVID-19) to 82.9% (during COVID-19). So more students failed the final exam but those who passed received on average higher grades. From personal discussions with students, we know that this decrease in passing the exam was at least partially caused by personal circumstances due to COVID-19. The increase in student satisfaction in the course and teacher performance on the other hand shows that students did not make the course or teacher responsible for having to take the resit exam.

Overall, we conclude that despite safety restrictions, high teaching quality with hands-on robotics experience for students could be achieved. We envision that some concepts developed and tested due to COVID-19 restrictions will even be maintained after the pandemic. In the future we envision to maintain video lectures for core and background material as well as instructional videos for tutorials and lab assignments as they allow students to study at their own pace and as staff gets to concentrate on activities to better support and challenge students, meeting their individual capabilities and needs. Still as teachers we are looking forward to more personal interaction with our students again without the need of keeping large distances to students that often generate awkward situations where we either have to speak to each other with an unnatural loud voice or communicate through IT infrastructure despite being in the same room. We also plan to come back to our flexible open teaching approach where students get invited to join our robot lab outside class to pursue their research ideas.

In the hope that our work on EDMO and the related course content will be beneficial for other teachers and students, we provide continuously updated material online including demonstrations of EDMO hardware and access to design files at https://www.maastrichtuniversity.nl/edmo.

Acknowledgment. We thank the many students who participated in this course and who made this course so enjoyable with their enthusiasm, commitment and creativity and who worked so hard despite the challenging COVID-19 conditions. We also thank Dieudonnée van de Willige for her kind support in creating and maintaining the EDMO website.

References

1. Bonwell, C.C., Eison, J.A.: Active Learning: Creating Excitement in the Classroom. ERIC (1991)
2. Felder, R.M., Brent, R.: Designing and teaching courses to satisfy the abet engineering criteria. J. Eng. Educ. **92**(1), 7–25 (2003). https://doi.org/10.1002/j.2168-9830.2003.tb00734.x
3. Wieman, C.: Science education in the 21st century–using the tools of science to teach science. In: Future of Higher Education, pp. 61–64 (2008)
4. Weimer, M.: Learner-Centered Teaching: Five Key Changes to Practice. Wiley, Hoboken (2002)
5. Möckel, R., Dahl, L., Christopher, S.M.: Interdisciplinary teaching with the versatile low-cost modular robotic platform EDMO. In: Moro, M., Alimisis, D., Iocchi, L. (eds.) Educational Robotics in the Context of the Maker Movement, EduRobotics 2018. Advances in Intelligent Systems and Computing, vol. 946, pp. 135–146. Springer, Cham (2019). https://doi.org/10.1007/978-3-030-18141-3_11
6. McCarthy, J.M.: Introduction to Theoretical Kinematics. MIT Press, Cambridge (1990)
7. Minorsky, N.: Directional stability of automatically steered bodies. J. Am. Soc. Naval Eng. **34**(2), 280–309 (1922). https://doi.org/10.1111/j.1559-3584.1922.tb04958.x
8. Ijspeert, A.J.: Central pattern generators for locomotion control in animals and robots: a review. Neural Netw. **21**(4), 642–653 (2008). https://doi.org/10.1016/j.neunet.2008.03.014
9. Garrison, D.R., Vaughan, N.D.: Blended Learning in Higher Education: Framework, Principles, and Guidelines. Wiley, Hoboken (2008)
10. Abeysekera, L., Dawson, P.: Motivation and cognitive load in the flipped classroom: definition, rationale and a call for research. High. Educ. Res. Dev. **34**(1), 1–14 (2015). https://doi.org/10.1080/07294360.2014.934336
11. Osguthorpe, R.T., Graham, C.R.: Blended learning environments: definitions and directions. Q. Rev. Distance Educ. **4**(3), 227–234 (2003)
12. Dabbagh, N., Kitsantas, A.: Personal learning environments, social media, and self-regulated learning: a natural formula for connecting formal and informal learning. Internet High. Educ. **15**(1), 3–8 (2012). https://doi.org/10.1016/j.iheduc.2011.06.002
13. Likert, R.: A technique for the measurement of attitudes. Arch. Psychol. **22**(140) (1932)

Distance Learning in the Era of COVID-19: Supporting Educational Robotics with Augmented Reality

Christina Pasalidou[1,2] and Nikolaos Fachantidis[1,2](✉)

[1] Department of Educational and Social Policy, University of Macedonia, Thessaloniki, Greece
{cpasalidou,nfachantidis}@uom.edu.gr

[2] Laboratory of Informatics and Robotic Applications in Education and Society, University of Macedonia, Thessaloniki, Greece

Abstract. Due to the Covid-19 pandemic, new challenges have emerged, bringing distance learning into the spotlight. In the context of distance learning and robotics education, we propose a system that utilizes an e-learning platform, an educational robot and mobile augmented reality technology. The current circumstances prompt this study to focus on those technologies to overcome the online distance learning deficits compared to in-person laboratory robotic activities implementation. This empirical study shows the preliminary results of the pilot implementation of this teaching approach. Combining AR technology with educational robotics is deemed an innovative and contemporary proposal for teaching students STEM concepts for online distance learning environments. An e-learning mobile AR educational robotics environment was designed for this study. In this paper, a qualitative pilot study is presented. Two groups of elementary school students participated in the study. Students' perceptions and behavior provided feedback regarding the feasibility and support of e-learning using an augmented reality environment at the educational robotics course. All of the students participated actively in the learning process, considering this type of teaching more pleasant and enjoyable, and in general, students' attitudes were positive. Future research should be made to examine the opportunities and challenges of implementing such a system in educational settings more in-depth.

Keywords: Covid-19 · E-learning · Educational robotics · Augmented reality

1 Introduction

This study aimed to investigate the implementation of an e-learning Augmented Reality robotics education system and the students' perceptions and behavior towards this system. Due to the Covid-19 pandemic, new challenges have emerged, bringing distance learning into the spotlight. Thus, an e-learning system for teaching robotics utilizing mobile Augmented Reality was designed for the purpose of this study. In the following sections of this paper the educational value of robotics, mobile augmented reality and e-learning are pointed out and the theoretical framework of the study is presented. In

M. Malvezzi et al. (Eds.): EDUROBOTICS 2021, SCI 982, pp. 39–51, 2021.
https://doi.org/10.1007/978-3-030-77022-8_4

research methodology, the design of the proposed system, including the e-learning platform, the educational robot and the mobile AR application, is also explained. Preliminary results of the pilot study and conclusions are mentioned and discussed.

For the past years, teaching has become more innovative and contemporary, using tools such as simulations, multimedia, mobile applications, augmented reality technology, distance learning platforms and new technologies in general, so as to provide students with more up-to-date teaching practices. According to Lytridis et al. [1], information and communication technologies (ICT) provide new means for content dissemination and user interaction, both with the content and the learning platforms. The extended use of the internet in general and the rise of technologies that utilize the internet have led to the introduction of new pedagogical approaches [2] that aim to provide students knowledge in an engaging and effective way. Thus, combining educational robotics, augmented reality and distance learning is deemed to be a promising system to achieve the goals and expectations of the 21st century rapidly changing educational settings.

The new challenges that have emerged in Education due to the Covid-19 pandemic, brought distance learning into the spotlight. In the context of distance learning, the robotic laboratory activities demand adaptations due to lack of equipment at student's home (e.g. robot or activity track) or other restrictions (e.g. limited space availability). Besides these limitations, online distance teaching demands an appropriate technical environment and support to be successfully used in educational settings [5, 7]. Potentially, all the previous issues can be an obstacle to students' smooth participation in distance learning activities and cause negative experiences. We propose a system that utilizes an e-learning platform, an educational robot and mobile augmented reality technology. The current era of COVID-19 circumstances prompt this study to focus on those technologies and if they are capable to overcome the online distance learning deficits compared to in-person laboratory robotic activities implementation. The research question, which this empirical study investigates through qualitative analysis, is if the students will easily get used to the specific e-learning environment, participate actively and experience enjoyable learning with a positive attitude. It also focuses on recording any issues that arise during the implementation of distance learning laboratory activities in the designed environment and the pilot implementation.

2 Related Work

2.1 E-learning and Covid-19 Pandemic

The Covid-19 pandemic brought a majority of challenges and changes in people's everyday lives. A lot of measures have been implemented to secure public health. In Education, online teaching was one of them. The vulnerability in school settings regarding the virus spread caused thousands of school closures [3]. Students, educators, and educational organizations all over the world were affected by the pandemic and the new circumstances [4].

According to Basilaia and Kvavadze [5], different countries worldwide have introduced various solutions during the pandemic to continue the education process, utilizing online libraries, TV broadcasts, guidelines, resources, video lectures and online channels.

In their research, Alea and his colleagues [6], studied teachers' awareness about Covid-19, their experiences from distance learning, their perceptions towards institutional readiness, as well as the challenges of conducting distance learning education in the Philippines. The results of their study showed that teachers might feel ready to implement distance teaching. However, they are concerned about the lack of facilities, equipment, and capacity building required for distance education [6].

From the students' perspective, distance education does not appear to be as motivating as face-to-face learning [7]. The absence of factors, such as the interaction with the teacher, the instant response and the traditional classroom socialization along with the technical difficulties that arouse, make e-learning even more challenging [7].

E-learning was the immediate measure to implement during Covid-19 quarantine. As a consequence, educational robotics courses were also conducted remotely [40]. While there have already been studies designing and using remote teaching and virtual learning for robotics [41–45], the current circumstances required the development of more e-learning practices for robotics education. However, there are some challenges that need to be addressed. Robotics usually requires using some high cost, special equipment that students do not have at home and simulation systems are not capable of providing students with the experience of working with a real robot [46]. In addition to that, due to enhancement of e-learning problems the learning performance of e-learning systems gets degraded [47]. For this reason, adaptive and effective approaches in solving practical and design problems need to be developed [46].

2.2 Educational Robotics in STE(A)M Education

Robotics can be a tool to motivate and enhance students' interest in learning STEM concepts [8, 9]. In educational robotics students build and program small robotic systems, designed especially for educational purposes [10]. According to Eguchi [11], the process of designing, constructing and programming autonomous robots helps students understand the technology and apply the knowledge learned in school in a meaningful and fascinating way. To be more specific, when dealing with educational robotics, students are put in a position which prompts them to use their knowledge from Science, Technology, Engineering and Mathematics (STEM) [12].

Educational robotics has a positive effect in students' STEM self-efficacy and interest in those fields of study, supporting experiential learning [13]. It provides learners with engaging and hands-on learning STEM experiences [14]. According to Barker and his colleagues [15], educational robots can be a useful tool in STEM education, prompting students to think like scientists and find solutions to problems presented. In educational robotics courses, students typically deal with authentic problems and use their prior knowledge as a basis for constructing further knowledge [9].

The development of social, as well as, cognitive skills is enhanced through educational robotics [16]. Students are put in situations where they have to apply critical thinking and problem solving [17]. In general, educational robotics could be characterized as a transformational means for education, supporting Science learning, computational thinking, coding and engineering which are the basic components of STEM learning in K-12 education [11].

While STEM curriculum has significantly evolved over the years, teachers hold the view that young learners should be equipped with the 21st century skills that include creativity, innovation, and entrepreneurship [33]. This can be achieved by adding arts and design to the equation, thus transforming STEM into STEAM education [34]. The inclusion of the Arts into STEM has led to the STEAM approach, where all five subjects are blended together for a topic of study [35], providing a well-rounded approach to education [36].

Following the principles of STEAM education, robotics in the classroom has the ability to support learning goals in the different subject areas [37]. Blending arts and humanities with technology and educational robotics helps students better understand the integration of STEM subjects with artistic, creative processes and design thinking [38]. According to Sullivan et al. [39], robotics is an engaging educational tool for teaching foundational STEAM concepts to young children.

The tangibility of robots and the excitement they bring into the classroom environment is considered conducive for learning [13]. Increasing students' engagement and interest in STEM [18] and STEAM concepts, educational robotics promotes active learning and improves the learners' experience [9], being in accordance with the demands of 21st century education.

2.3 Robotics and Augmented Reality

Augmented Reality (AR) is considered as the enhanced environment where the Virtual and the Real world meet [19]. Augmented Reality has been a growing field of study in recent years. A variety of AR applications have been designed and created, mainly utilizing mobile devices [20].

In the field of Education, several studies have been conducted, making use of an Augmented or Mixed Reality system and a robot, with the aim to achieve learning goals [21] by enhancing student motivation [22]. According to Chang and his colleagues [22], integrating robots into Mixed Reality environments improves students' authentic learning experiences.

Mobile devices, such as tablets, are widely used in Augmented Reality environments. In Frank et al. study [23], users manipulated the virtual objects through the tablets, programming in this way the robot to head towards the physical objects, pick them up and change their location.

Robots can be programmed through the smart mobile devices with the aid of Augmented Reality taking advantage of the possibilities it provides. In Gradmann et al.'s research [24], the AR technology made the enhanced view of the real objects and the preview of the robot's moves possible for the users to see. This system required no specialized programming knowledge and was quite adaptive and easy to use [24].

In educational robotics, Xefteris and Palaigeorgiou [25] combined three dimensional, tangible and virtually enhanced learning environments. They augmented the users' view and created treasure hunt activities. The robots used were Lego Mindstorms EV3. Results from their study showed that the learning process becomes more engaging, participatory and interesting, while activating students, promoting collaboration and developing their computational thinking [25].

3 System Design

For the purpose of this study, an e-learning Augmented Reality system/environment was designed for teaching robotics to elementary school students. The e-learning platform used was Google Meet. It is a free, simple and easy to use platform with many good reviews. All participant students had Google account and so it was effortless for them to join a Google Meet call. Taking those advantages into consideration, we were led to choosing this platform for the distance learning activities. The researchers shared the link of the online lessons in each group via email and set the dates and time of each lesson. Students used the link to join the online meeting and the educator granted their access. All participants were asked to join the online class 15 min before each lesson in order to check if everything is working properly in the e-learning platform and avoid connectivity problems that could lead to the delay of the lesson. Both students and the educator had their cameras and microphones on during the lessons. Thus, students could interact with the teacher and their peers, as well.

The educational robot that was used was the Edison robot. It is a small mobile robot, designed to be used in educational settings in order to teach children programming. In contrast with other educational robots, Edison robot is very affordable and easy to use, having different programming environments for each age group of students. The visual programming environment with drag and drop commands is widely used in classrooms. An additional positive feature of the Edison robot is that it is designed in a way that Lego building bricks can be assembled on its top surface as well as on the sides (Fig. 1) with a functional or/and aesthetic aim. Edison speed was chosen so that the augmented reality system has the best reliability in monitoring the pointer placed on it.

Fig. 1. The Edison robot with the addition of Lego bricks

The pilot study included three lessons of two didactical hours each. The lessons were designed in a way that they utilized different objects and devices every time and had escalating difficulty regarding the robot programming. The content was adjusted for each age group and aligned with the principles of STEAM education.

The first lesson was about "Geometry in Art". It was an introductory lesson where students became familiarized with the e-learning platform. They practiced how to navigate in the platform, how to change tabs and how to share their screen. They also saw where the computer camera was and how they could show there their robot. In this educational robotics lesson, students were asked to program their robot to make a drawing with a variety of geometric shapes. Students achieved that using objects that they already had at home, such as colorful markers and tape in order to build their robot and equip it with the necessary means to be able to complete the given tasks. As mentioned, the Edison robot offers the opportunity to adjust lego bricks both on its surface and on the sides. This feature prompt the researchers to add activities asking students to be inspired and create a component made of lego bricks they owned. This component would replace the tape and would be used to keep the markers steady.

The second lesson focused on "Keeping the Environment Clean". After having an online conversation and learning about environmental pollution, its causes and solutions, students were familiarized with more commands in the programming environment of the Edison robot. Consequently, the tasks for the robot in the activity sheet were more difficult and demanding. In this case, students were provided with lego building bricks (non-electronic) from the educational robotics package Lego WeDo 2.0 and the Lego Mindstorms EV3 and given specific instructions on how to build their robot. The additional component was a caliper that made the Edison robot functional and able to support students finish their mission, cleaning the environment from the rubbish.

The third lesson's theme was Automatic Navigation. This lesson utilized mobile devices and an educational Augmented Reality application which was designed for the purpose of this study. To be more specific, Unity 3D and Vuforia were the software used to develop this AR application. Apart from the e-learning platform, the Edison robot and the programming environment which students had become quite familiar with, they got to know Augmented Reality technology and its mode of operation. Firstly, students were instructed to download the AR app on their Android mobile device, either smartphone or tablet. In order for the application to work, students were provided with printed markers/target images. After opening the educational AR app, they were asked to point the camera at the target images. Onto the one image, a whole city was overlaid and appeared in the users' real environment, while at the same time the second target image triggered the application to show the vehicle that the Edison robot would transform to (Fig. 2). For this reason, the marker with the city would be placed on the floor in order to have more space and the marker with the vehicle would be taped on the robot. Students were prompt to program their robot to navigate in the AR city. To accomplish that, students would have to experiment, estimate and use trial and error until they succeed. AR' role in this case was to provide the context where students would work and program their robot accordingly, like having a robotics mat, but virtual and more interactive, pointing the places the robot has to go and showing 3D animations when it reaches the desired destination.

4 Empirical Study

In order to evaluate the proposal of this work, a pilot study has been conducted. To be more specific, the aim of this study was to test the applicability of an e-learning Augmented Reality educational Robotics course, consisted of three lessons. During the lessons, students' behavior, attitude towards the proposed system, possible difficulties,

Fig. 2. Snapshots from the AR application and view of the two markers/trigger images (on the Edison robot and on the table).

limitations and also positive aspects would be observed by the researchers in order to give feedback regarding the modifications that should be made to proceed in further research and experimental studies in a wider range of students.

4.1 Participants

The participants of this pilot study were 10 Elementary School students. Three of them were female and seven were male. The students were separated into two groups, regarding their age. The first group consisted of 5 students between 6–8 years old and the second group consisted of 5 students between 9–12 years old. The participants had no prior knowledge of using and programming the Edison robot. Two of them (20%) had used an e-learning platform for school before, while it was the first time for the rest of the participants.

4.2 Implementation and Data Collection

The robotics course included three 2-h lessons for each group of students. Each lesson differed in terms of difficulty and content. In the first lesson students utilized objects they owned and had in their home, in the second lesson they used lego bricks from educational robotics packages along with building instructions and in the third lesson they programmed their robot to navigate in an augmented reality setting, using a mobile device and markers/target images.

Data were collected using an observation sheet. The observations during the e-learning activities were made by two observers, specialized in educational robotics and new technologies in the field of Education. Based on Junker's field observer roles [26] and Gold's analysis [27] on their characteristics, the one researcher had the role of the *complete observer*, maintaining some distance from the participants, being detached and not interacting with them. The second observer's role would be the *participant as observer*, being involved in the activities, communicating with the participants and informing them from the beginning about their observational role. Participants were observed during the whole time of the online classes. Observations are quite helpful in order to detect what is important to the participants, how much time is spent on various activities, if there are non-verbal expressions of feelings [29, 30]. For this reason, both observers took notes, focusing on students' comments, expressions, reactions, queries, difficulties they encountered, understanding of lesson content, participation and time they spent on each task. Through observation it is possible to ascertain whether peoples' actions align with their claims or not [28].

At the end of each lesson, one researcher would ask students about their impressions of the lesson. After the end of the course, participants were questioned regarding their experience about these distance learning lessons, robotics teaching, and mobile augmented reality. They were also asked to compare the three lessons, evaluate them, declare the lesson they enjoyed the most, and justify their answers.

4.3 Results

From the researchers' observations the following results are worth mentioning.

Robot Programming

After discussing with the participants, it was stated that none of them had used an Edison robot before. In addition to that, they had never seen the programming environment of Edison robot. Students seemed to enjoy the lesson and they carried out the programming tasks successfully. The drag 'n' drop programming commands (blocks) made the process easier to the students that participated in the study.

Augmented Reality

Only one of the participants had previously used an AR application, but it was not related with robotics education. The rest of the participants were not aware of the Augmented Reality technology and had never used an educational AR application in the past. They seemed to understand the technology of AR and the relation between the markers/target images, the camera of the mobile device and the virtual 3D models which are superimposed into the real environment.

In the beginning of the AR activity some students encountered some technical problems with the installation of the educational AR application, but they were quickly solved. The majority seemed to enjoy the lesson with the AR technology the most. To be more specific, to the question "Would you like in the future the educational robotics' lessons to include AR activities?" all participants were positive whilst a student replied "Yes, it would be really nice to have AR activities while learning robotics, because it is something different, something new that we did not have before".

Observers pointed out that although the AR activity was carried out successfully by the participants, it would be even more effective to have students working in pairs and changing roles. To be more specific, the one student could hold the mobile device and view the virtual content while the other one would program the robot in the programming environment, based on his/her teammate's instructions for the robot's navigation in the AR city.

E-learning system

The e-learning system was quite new for the students in general. The participants of the second group (9–12 years old) were more familiarized both with the mobile devices and with the computer and the e-learning platform. The younger students, however, encountered some difficulties, mostly when they had to change tabs in their browser and then return to the e-learning platform. They all understood how the platform worked after its buttons and functionalities were explained to them in detail.

From the first lesson they participated actively in the teaching process, having their cameras and microphones on the whole time. They raised hands, asked questions, expressed their opinions and answered the educator's queries. Additionally, students were presenting and showing their robot and the way it behaved when their program was running by turning the camera of their computer. They also pointed their robot or the mobile device screen to the computer camera when they faced some difficulty so that the educator could assist them in a more efficient way.

Since all participants had a good internet connection, there were not connectivity issues, apart from one time when the student participated in the lesson using a mobile device for a few minutes. In the e-learning lesson, the educator needed time to make sure that all participants are on the same spot and no one is left behind. Additional limitations that were evident though this study were the lack of in-person interaction both with the educator and the students, as well as the limited collaboration between the students.

Concluding, using the proposed system was a positive experience for the participants. Students agreed that with the mobile devices and the Augmented Reality technology the robotics lesson becomes more enjoyable and pleasant compared to the teaching practices they have been used to. Learning through trial and error helped students program their Edison robot on their own and reach the teaching objectives. Overall, as they described it, it was a constructive and enjoyable experience.

5 Discussion and Conclusion

During the Covid-19 pandemic, educators, students and educational organizations have been coping and adjusting to the distance learning education [6]. A number of recent studies have researched the challenges and opportunities associated with e-learning during pandemics [4]. The sudden shift from traditional classrooms and teaching in person to online learning has resulted in a completely different learning experience for students [7]. However, in order to align with the appeared changes there is an urgent need for a radical and effective educational reform [31].

In this paper, a pilot study has been conducted in order to assess the new learning environment designed and proposed, consisting of Educational Robotics and mobile Augmented Reality technology in an e-learning system. Systems utilizing augmented or mixed reality technology and robots can have positive outcomes, affecting students' motivation and learning performance [22]. Qualitative results of this pilot study indicated that students easily got used to the e-learning platform as the lessons came by. All of the students participated actively in the teaching process. Preliminary results also suggest that mobile devices and Augmented Reality technology make learning more pleasant and enjoyable for the students. AR provided a more interactive way to assist students in programming their robot, showing a map with virtual elements where the robot navigated as a 3D model of a vehicle. In general, students' attitudes were positive, concluding that AR combined with educational robotics can be a valuable educational tool that students would like to use in the future.

This study aimed to provide students with a unique learning experience, where they could compare the online lessons they participated in. Under the circumstances of Covid-19 pandemic, in-person teaching cannot be implemented. In this case, e-learning is the solution to continue educating students remotely. This pilot study was designed to show the feasibility and the challenges of using an e-learning Augmented Reality system for robotics education. The results, along with the data collected from the observations showed the students' perceptions and the limitations of this approach, such as the need of equipment, the technical issues that may appear, the limited collaboration between the students, the lack of physical interaction and the difficulty in monitoring where each student is during the lesson so that no one falls behind.

This study was the first stage of our research, pointing out the improvements that could be made in order the system to be more effective and easier to use. In future studies, we intent to investigate the implementation of the proposed system in a wider scale of students and combine both qualitative and quantitative research. More research is essential so as to examine the challenges of utilizing e-learning [7] in accordance with educational robotics and AR technology. AR can assist and improve the teaching of robotics education, leading to a better understanding, further discussion, and correction of the robot code [32]. Meanwhile, we should have in mind that online teaching demands an appropriate technical environment and support to be successfully used in educational settings [5].

Acknowledgments. This research is co-financed by Greece and the European Union (European Social Fund- ESF) through the Operational Programme "MATCHING FUNDS - Science Technology Innovation Mathematics Engineering for the Young (STIMEY)".

References

1. Lytridis, C., Tsinakos, A., Kazanidis, I.: ARTutor—an augmented reality platform for interactive distance learning. Education Sciences **8**(1), 6 (2018)
2. Alzahrani, N.M.: Augmented reality: a systematic review of its benefits and challenges in e-learning contexts. Applied Sciences **10**(16), 5660 (2020)
3. Toquero, C.M.: Challenges and opportunities for higher education amid the COVID-19 pandemic: the Philippine context. Pedagogical Res. **5**(4), em0063 (2020)
4. Mailizar, A.A., Maulina, S., Bruce, S.: Secondary school mathematics teachers' views on e-learning implementation barriers during the Covid-19 pandemic: the case of Indonesia. Eurasia J. Math. Sci. Technol. Educ. **16**(7), em1860 (2020)
5. Basilaia, G., Kvavadze, D.: Transition to online education in schools during a SARS-CoV-2 coronavirus (COVID-19) pandemic in Georgia. Pedagogical Res. **5**(4), em0060 (2020)

6. Alea, L.A., Fabrea, M.F., Roldan, R.D.A., Farooqi, A.Z.: Teachers' Covid-19 awareness, distance learning education experiences and perceptions towards institutional readiness and challenges. International Journal of Learning, Teaching and Educational Research **19**(6), 127–144 (2020)
7. Adnan, M., Anwar, K.: Online learning amid the COVID-19 pandemic: students' perspectives. Online Submission **2**(1), 45–51 (2020)
8. Cuellar, F., Penaloza, C., Garret, P., Olivo, D., Mejia, M., Valdez, N., Mija, A.: Robotics education initiative for analyzing learning and child-parent interaction. In: 2014 IEEE Frontiers in Education Conference (FIE) Proceedings, pp. 1–6. IEEE (2014)
9. Anwar, S., Bascou, N.A., Menekse, M., Kardgar, A.: A systematic review of studies on educational robotics. J. Pre-College Eng. Educ. Res. (J-PEER) **9**(2), 2 (2019)
10. Stergiopoulou, M., Karatrantou, A., Panagiotakopoulos, C.: Educational robotics and STEM education in primary education: a pilot study using the H&S electronic systems platform. In: International Conference EduRobotics 2016, pp. 88–103. Springer, Cham (2016)
11. Eguchi, A.: Robotics as a learning tool for educational transformation. In: Proceedings of 4th International Workshop Teaching Robotics, Teaching with Robotics and 5th International Conference Robotics in Education, Padova, pp. 27–34 (2014)
12. Alimisis, D., Karatrantou, A., Tachos, N.: Technical school students design and develop robotic gear-based constructions for the transmission of motion. In: Eurologo Conference "Digital Tools for Life Long Learning", Warsaw, pp. 76–86(2005)
13. Mead, R.A., Thomas, S.L., Weinberg, J.B.: From grade school to grad school: an integrated STEM pipeline model through robotics. In: Robots in K-12 Education: A New Technology for Learning, pp. 302–325. IGI Global (2012)
14. Mohr-Schroeder, M.J., Jackson, C., Miller, M., Walcott, B., Little, D.L., Speler, L., Schroeder, D.C.: Developing middle school students' interests in STEM via summer learning experiences: see blue STEM camp. School Science and Mathematics **114**(6), 291–301 (2014)
15. Barker, S.B., Nugent, G., Grandgenett, N., Adamchuk, I.V.: Robots in K-12 Education: A New Technology for Learning. Information Science Reference. IGI Global, New York (2012)
16. Ioannou, A., Makridou, E.: Exploring the potentials of educational robotics in the development of computational thinking: a summary of current research and practical proposal for future work. Education and Information Technologies **23**(6), 2531–2544 (2018)
17. Bers, M.U., Flannery, L., Kazakoff, E.R., Sullivan, A.: Computational thinking and tinkering: exploration of an early childhood robotics curriculum. Comput. Educ. **72**, 145–157 (2014)
18. Kim, C., Kim, D., Yuan, J., Hill, R.B., Doshi, P., Thai, C.N.: Robotics to promote elementary education pre-service teachers' STEM engagement, learning, and teaching. Comput. Educ. **91**(C), 14–31 (2015)
19. Milgram, P., Kishino, F.: A taxonomy of mixed reality visual displays. IEICE Transactions on Information and Systems **77**(12), 1321–1329 (1994)
20. Kurubacak, G., Altinpulluk, H.: Mobile Technologies and Augmented Reality in Open Education. IGI Global, Hershey, PA (2017)
21. Chen, G.D., Chang, C.W.: A task-based role-playing game with educational robots for learning language. In: International Conference on Technologies for E-Learning and Digital Entertainment, pp. 483–488. Springer, Heidelberg (2009)
22. Chang, C.W., Lee, J.H., Wang, C.Y., Chen, G.D.: Improving the authentic learning experience by integrating robots into the mixed-reality environment. Comput. Educ. **55**(4), 1572–1578 (2010)
23. Frank, J.A., Moorhead, M., Kapila, V.: Realizing mixed-reality environments with tablets for intuitive human-robot collaboration for object manipulation tasks. In: 2016 25th IEEE International Symposium on Robot and Human Interactive Communication (RO-MAN), pp. 302–307. IEEE (2016)

24. Gradmann, M., Orendt, E.M., Schmidt, E., Schweizer, S., Henrich, D.: Augmented reality robot operation interface with Google Tango. In: ISR 2018 50th International Symposium on Robotics, pp. 1–8. VDE (2018)
25. Xefteris, S., Palaigeorgiou, G.: Mixing educational robotics, tangibles and mixed reality environments for the interdisciplinary learning of geography and history. International Journal of Engineering Pedagogy (iJEP) **9**(2), 82–98 (2019)
26. Junker, B.: Some suggestions for the design of field work learning experiences. Cases on Field Work. Chicago University Press, Chicago III–A (1952)
27. Gold, R.L.: Roles in sociological field observations. Soc. Forces **36**(3), 217–223 (1958)
28. Mulhall, A.: In the field: notes on observation in qualitative research. J. Adv. Nurs. **41**(3), 306–313 (2003)
29. Kawulich, B.: Collecting data through observation. Doing social research: A global context, pp. 150–160 (2012)
30. Schmuck, R.A.: Practical Action Research for Change. Skylight Training and Publishing, Arlington Heights, IL (1997)
31. Eguchi, A.: Bringing robotics in classrooms. Robot. STEM Educ. 3–31 (2017)
32. Cheli, M., Sinapov, J., Danahy, E.E., Rogers, C.: Towards an augmented reality framework for k-12 robotics education. In: Proceedings of the 1st International Workshop on Virtual, Augmented, and Mixed Reality for HRI (VAM-HRI) (2018)
33. Khine, M., Areepattamannil, S.: Steam Education. Springer, Cham (2019)
34. Liao, C.: From interdisciplinary to transdisciplinary: an arts-integrated approach to STEAM education. Art Educ. **69**(6), 44–49 (2016)
35. Smith O.: There is an art to teaching science in the 21st century. In: Ge, X., Ifenthaler, D., Spector, J. (eds.) Emerging Technologies for STEAM Education. Educational Communications and Technology: Issues and Innovations, pp. 81–92. Springer, Cham (2015)
36. Rolling Jr., J.H.: Reinventing the STEAM engine for art + design education. Art Educ. **69**(4), 4–7 (2016)
37. Hamner, E., Cross, J.: Arts and Bots: Techniques for distributing a STEAM robotics program through K-12 classrooms. In 2013 IEEE Integrated STEM Education Conference (ISEC), pp. 1–5. IEEE (2013)
38. Damaševičius, R., Maskeliūnas R., Blažauskas, T.: Faster pedagogical framework for steam education based on educational robotics. Int. J. Eng. Technol. **7**(2.28), 138–142 (2018)
39. Sullivan, A., Strawhacker, A., Bers, M.U.: Dancing, drawing, and dramatic robots: integrating robotics and the arts to teach foundational STEAM concepts to young children. In: Robotics in STEM Education, pp. 231–260. Springer, Cham (2017)
40. Gomes, A.S.A., Da Silva, J.F., Teixeira, L.R.D.L.: Educational robotics in times of pandemic: challenges and possibilities. In: 2020 Latin American Robotics Symposium (LARS), 2020 Brazilian Symposium on Robotics (SBR) and 2020 Workshop on Robotics in Education (WRE), pp. 1–5. IEEE (2020)
41. Casini, M., Garulli, A., Giannitrapani, A., Vicino, A.: A remote lab for experiments with a team of mobile robots. Sensors **14**(9), 16486–16507 (2014)
42. Andreev, V., Kuvshinov, S., Pryanichnikov, V., Poduraev, Y.: Education on the basis of virtual learning robotics laboratory and group-controlled robots. In: Procedia Engineering, vol. 69, pp. 35–40. Elsevier (2014)
43. Abreu, P., Barbosa, M.R., Lopes, A.M.: Robotics virtual lab based on off-line robot programming software. In: 2013 2nd Experiment@ International Conference, pp. 109–113. IEEE (2013)
44. Jara, C.A., Candelas, F.A., Torres, F.: Virtual and remote laboratory for robotics e-learning. In: Computer Aided Chemical Engineering, vol. 25, pp. 1193–1198. Elsevier (2008)
45. Candelas, F.A., Puente, S.T., Torres, F., Ortiz, F.G., Gil, P., Pomares, J.: A virtual laboratory for teaching robotics. Complexity **1**, 10–11 (2003)

46. Oleinikov, V., Klimov, K., Rogachev, A., Kapytov, D., Budanov, V., Shtcherbov, R., Slabukha, N., Buryak, A.: The remote robotics education approach, its benefits and difficulties, and the review of distance accessed robotics training ground. In: Interactive Robotics Training Ground, pp. 1–11 (2020)
47. Khamparia, A., Pandey, B.: Association of learning styles with different e-learning problems: a systematic review and classification. Education and Information Technologies **25**(2), 1303–1331 (2020)

Mendieta, One Robot Per School: Multi-user Robot for Technology Education

Gonzalo Zabala[1](✉), Ricardo Morán[1,2], and Matías Teragni[1]

[1] Universidad Abierta Interamericana, Centro de Altos Estudios en Tecnología Informática, CABA, Buenos Aires, Argentina
{gonzalo.zabala,ricardo.moran,matias.teragni}@uai.edu.ar
[2] Comisión de Investigaciones Científicas, La Plata, Buenos Aires, Argentina

Abstract. In recent years, the use of physical computing devices in Argentinian schools has been growing steadily. The insertion of robotics in the classroom allows addressing a set of technology-related subjects, such as mechanics, electronics, and programming. It is no coincidence that the Argentine educational system has incorporated these subjects into its curricular design, as well as the introduction of new technical schools aimed at programming and robotics. However, the cost of the equipment, the difficulty of maintenance, and the technical knowledge required to be able to properly use this resource in an educational context are limiting factors that hinder the potential to reach all schools. For these reasons, a project has been designed to overcome these difficulties: a low-cost autonomous robot controlled by an Arduino Nano, with a web server built into an Orange Pi computer, which allows programming by multiple users at the same time, queuing the execution orders in such a way that with a single device an entire class can work seamlessly. This allows the introduction of robotics in a school with a cost of less than USD 180. Additionally, all the development of both hardware and software is open source, offering the community a framework for building other robots with the same architecture.

Keywords: Technology education · Robotics · Web server · Arduino · Orange Pi

1 Introduction

Since 2010, the introduction of robots in Argentine classrooms as a didactic resource has encouraged an increasing interest in incorporating the discipline at the curricular level along with programming (Ripani 2017). We can mention, among others, the project "Todos a la robótica" in the province of San Luis (Munizaga 2013), which from 2011 to 2014 incorporated material, training and a set of activities in all its primary schools. Another project worthy of mention is "Aprender conectados" (primarily called Escuelas del Futuro 2017) that reached about 3,000 schools throughout the country, both at primary and secondary level, including mechanics, electronics and robotics materials, as well as training and activities for use in the classroom (Ministerio de Educación – Presidencia de la Nación 2017). In the Province of Buenos Aires, we find the Plan Provincial de Robótica

M. Malvezzi et al. (Eds.): EDUROBOTICS 2021, SCI 982, pp. 52–63, 2021.
https://doi.org/10.1007/978-3-030-77022-8_5

Educativa (2018) that also reaches primary and secondary schools with different types of kits (Banchoff Tzancoff et al. 2019). Finally, in 2018 the Ministry of National Education defined for the first time learning objectives for compulsory education in programming and robotics, to start the path of their formal inclusion in the teaching and learning proposals in the schools of Argentina (Ripani 2017).

Even though there is a growing interest in the discipline, the sustainability of the activity over time is difficult to achieve. Physical materials suffer natural wear and tear in their use (which does not happen, for example, with computers or software), in addition to having an accelerated level of obsolescence. On the other hand, public policies related to the insertion of technology in schools are modified in each administration whenever there is a change of party in government. These experiences lead us to conclude that simply purchasing kits and training teachers is not enough: long-term support is necessary in order to ensure financial resources to periodically renew the material and activities.

In this paper we present "Mendieta, one robot per school". The objective of this project is to facilitate the insertion of robotics in secondary schools and its sustainability over time.

Given the aforementioned problems, this project should meet the following requirements:

- The development of the robot must be very low cost and highly reusable and upgradeable.
- The construction of the robot must be simple enough that can be implemented directly at the school.
- The hardware and software used must be fully open[1] to build a community that would be able to maintain and create new models.
- Building a community of teachers and educators is essential to provide activities that other teachers can use and extend.

Regarding low-cost solutions for educational robotics, various proposals can be found in the literature (Aroca et al. 2012; Darrah et al. 2018; Junior et al. 2013; Lopes Filho et al. 2011; López-Rodríguez and Cuesta 2016; Rubenstein et al. 2012; Saleiro et al. 2013). However, all these proposals imply the need for a robot per group of students and, in general, the usage of external devices such as smartphones. Additionally, the hardware they propose is too complex to be built by teachers without sufficient knowledge of electronics. To further reduce the cost, and meet the aforementioned objectives, Mendieta has been designed as a robot with a web server hosted on it, using a software system that allows for concurrent access to the programming tools using any modern web browser. In this way, with a single robot (or with a small number of them) the entire class can work simultaneously.

For the robot electronics, we have selected low-cost, standardized, easy to assemble, and easily available components. Regarding the robot structure, models are provided in STL format that can be printed on any of the cheapest 3D printers. Both electronics and 3D design have broad Creative Commons licenses.

[1] All the source code and design files will be published on https://github.com/MendietaProject.

In addition to the robotics kit, a website will be developed where teachers can consult activities classified by level of difficulty and associated subject. They will also be able to propose their own activities from a template form, which will then be moderated by our group before its final publication.

In the following sections we will describe the technical challenges and design decisions behind Mendieta. In the first section we will make an analysis of various educational robotics hardware architectures, ending with a description of the one chosen for Mendieta. And in the second part we will focus on the software architecture, both on the robot's firmware and on the programming platform mounted on the web server.

2 Hardware Architecture

In order to reduce the cost of the materials there are several things to consider. On the one hand, robotics equipment is not in permanent use in classrooms. Therefore, the robotics kit must be designed in a way that allows it to be shared between classes. This leaves out kits that involve construction processes that exceed the duration of a single lesson, otherwise the equipment can only be used by one class instead of being shared with the whole school.

At the same time, the kit must have the versatility to provide the students challenges of diverse levels of complexity. To that end, several low-cost robotic kits have already been proposed that meet these requirements. One of them is Cellbots (Aroca et al. 2012), which uses a smartphone as its processing unit. It proposes the use of the microphone input to control the motors and to read external sensors, as well as the use of other phone sensors to enrich the information about the robot surroundings. Another similar proposal is the Andruino - A1 (López-Rodríguez and Cuesta 2016), which includes an Arduino Uno as the device responsible for connecting sensors to an Android phone. In both cases, the model is still one robot per team, although the form of communication proposed in this last project would allow a similar use to the one offered by Mendieta. However, both projects assume the students would use their own smartphones, so they do not include it when estimating the final cost for each kit. And forcing students to use their own smartphones can be problematic, given the well-known risk of accidents involving mobile robots.

Other options such as those found in (Junior et al. 2013; Lopes Filho et al. 2011; Rubenstein et al. 2012) are harder to build by inexperienced users, and maintain the model of one robot per group of students.

For these reasons, we understand that the least possible amount of materials in a course is that of a single robot. With this objective, the proposed robot not only has the responsibility of coordinating its sensors and actuators based on the programs developed by the students, but it must also be able to receive, interpret and coordinate the simultaneous work of said students. To allow concurrent access to write, compile and execute programs, students use their own computers or cell phones to access a web page hosted on the robot, acting as local server, and through which they can program, debug, request the execution, and monitor the device status.

Since the established requirements exceed the limited capabilities of the most common microcontrollers used in educational robotics, there are two alternatives: either use specialized and expensive hardware, or complement the microcontroller with a microprocessor that has the necessary computing power and capabilities to function as a web server. **It is important to note that when we refer to both microcontrollers and microprocessors, we refer to integrated boards that complement the corresponding chips with the electronics and firmware necessary for an application without the need for complex technical knowledge**.

Popular and relatively low-cost microprocessors that have access to sensors and actuators already exist, but most do not include an analog-digital converter out of the box. Therefore, if you want to read analog sensors, such as light, humidity, distance, etc., you must add additional hardware and lose a significant amount of the ports that are available for sensors. Furthermore, this solution involves more complex electronic development.

Based on the above, the proposed architecture has two processing units. On the one hand, a microprocessor responsible for hosting the web server that lets teachers to monitor and configure the robot, as well as allowing students to program, debug and execute code. And on the other hand, a microcontroller in charge of executing the students' program, accessing the connected actuators and sensors.

There are multiple hardware options that can meet the established requirements, but given the nature of this project, which seeks to be open-software and open-hardware, the ease of access to the components required for assembly is an important factor to consider. The use of standard components in the educational and maker community worldwide was prioritized. Therefore, an Arduino was chosen as the microcontroller, being an open platform, low cost and widely accepted by the community, and an Orange Pi as the microprocessor, which has wide availability and documentation.

In particular, regarding the microcontroller, the Arduino Nano (Arduino Nano | Arduino Official Store, s/f) was chosen because it is one of the versions of the board most used in Argentina today, and because it can be replaced by other versions with a minimal work if necessary. It is less expensive and consumes less power than Arduino Uno, offering similar features, which are sufficient for this project.

Regarding the microprocessor, the Orange Pi Zero (orange pi zero - Orangepi, s/f) was selected mainly due to its low power consumption. Despite being less popular than Raspberry Pi, it has more processing power for a similar cost, allowing for better performance. However, a Raspberry Pi Zero W could be used instead, albeit at a lower performance.

So far, the minimum base of the hardware architecture has been defined. The objective is that different robot proposals could be developed based on this framework. Next, we will describe a first prototype which we will use to make the first experiences in the classroom.

Regarding the sensors, with just two types of sensors we can cover multiple different use cases: the ultrasonic sensor HC-SR04 (Freaks 2016) and the TCRT5000 module (Semiconductors 2009). Both sensors are low cost and have a strong market presence, making them an ideal choice for a robotics kit. We included one HC-SR04 to allow measuring distances with a range of 2 to 400 cm and an accuracy in the order of 3 mm. And we added two TCRT5000 modules, which are reflective optical sensors widely used for line and obstacle detection. The TCRT5000 modules have another advantage in that they already include the electronics required to connect with the Arduino.

Regarding the movement, two direct current motors with a built-in reduction box were selected (Motor Yellow 3-12VDC 2 Flats Shaft, s/f). These are also popular in the maker community. In order to synchronize the speed of each motor, an optical encoder is used. We decided to use an opto-switch fc-03 (MOCH22A Interrupter Datasheet pdf - OPTO Interrupter. Equivalent, Catalog, s/f) based on an LM393 comparator. The motors are controlled by the module based on the L298N driver (Handson Technology, s/f).

For power supply we decided to use two rechargeable lithium ion 14500 batteries of 3.7 V each, which have the same size as traditional AA batteries, thus being able to use a conventional battery holder. We also incorporated the TP4056 module (NanJing Top Power ASIC Corp., s/f) that allows to recharge them with an input voltage of 4.5 to 5.V and a miniUSB plug like that of traditional cell phone chargers.

Regarding the visualization of the robot data, in addition to the software tools that will be shown later, we added a 16 × 2 liquid crystal display compatible with the Hitachi HD44780 driver (Hitachi, s/f).

Finally, the mechanical structure of this first prototype is still in a design stage. The objective is to make a minimum amount of parts that can be obtained from the cheapest 3D printers on the market. It is for this reason that none of them will exceed 14 cm in any of its dimensions, since the printing beds have a limit of 15 cm in the economic models. To simplify the printing process and make the components easy to locate, the design was done in two layers. In the image below, the items in gray will be printed while the other components are placed to show the final structure of the robot (Fig. 1).

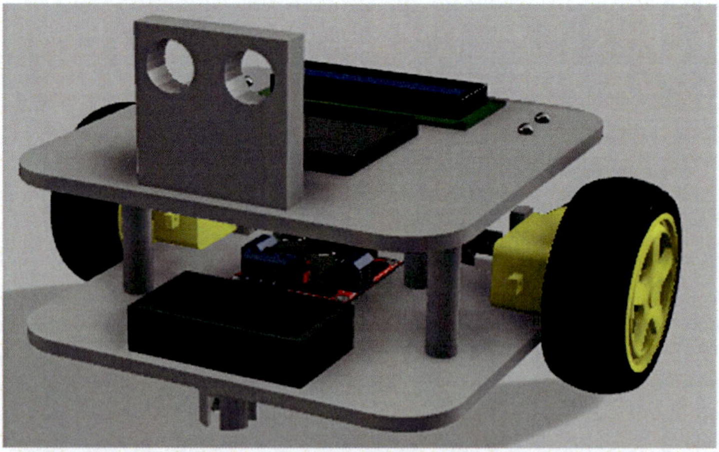

Fig. 1. 3D design of the first prototype based on the proposed architecture

We believe this design would benefit the sustainability of the project due to the minimum coupling between the hardware components. However, changes in the selected components have the potential to disrupt the hardware structure in ways that might require some changes in the design. We believe, though, that the open-source nature of the project and the flexibility of the design should allow for those changes to be performed without too much trouble.

3 Software Architecture

The inclusion of a web server in the robot itself is a proposal that was already present in the world of educational robotics. In (Saleiro et al. 2013) the objective is to enable the programming of the robot from any Wi-Fi device through a browser with javascript, and without installing any additional software. The use of the Blockly framework as a programming tool is also proposed. In other cases (Erdemir et al. 2015), the server has the function of allowing the use of the robot remotely. But in none of the examples found, the objective is to make a concurrent use of the robot by the students.

The following architecture diagram shows the different components of the system and how they interact to allow all the students in the classroom to simultaneously program the robot (Fig. 2).

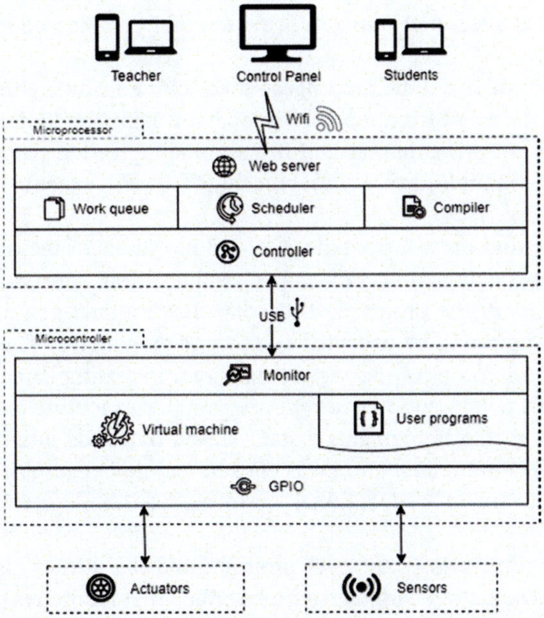

Fig. 2. Mendieta's software architecture diagram

Both students and teachers will be able to access the device through the web browser (already present on users' devices). In order to do this, the robot's microprocessor will expose a web server that will grant access to a graphical user interface including both the configuration and administration of the programming queue, as well as the integrated development environment from where students can program the robot.

This design will allow the robot to be programmed wirelessly (through the Wi-Fi network) and using any type of device (notebook, desktop computer, cell phone, tablet, etc.) without requiring users to install anything. Optionally, a control panel can also be connected to allow the entire classroom to view the queue of pending programs.

The scheduler module is responsible for deciding when each program will be executed. To do this, it uses the work queue, which stores each of the programs sent by the students. By default, the order of arrival will be respected, but if the teacher so wishes, the order of execution of the assignments can be manually modified, as well as eliminating programs from the queue. Whenever a program is ready to be executed a message will appear in the students' device notifying them that it is their turn to use the robot. At that time, students will have a fixed amount of time to test their program and evaluate how well it performs the assigned task.

The compiler module is responsible for generating the instructions that the microcontroller will execute as well as ensuring that the programs added to the queue are syntactically valid. The compiler can be invoked by any user connected to the robot at any time, this way the students can verify their programs before submitting them to the work queue.

The controller module will be responsible for the communication with the microcontroller. This includes managing the program execution and allowing both control of the robot as well as receiving status information that can then be made available to the web server.

The microcontroller will be preprogrammed with a firmware that allows interactive programming of the robot from the microprocessor using the Serial port. This firmware consists of a monitor program that will be responsible for the communication protocol with the controller module, and a virtual machine that will execute the programs written by the user.

The monitor program will periodically send the status of the robot to the controller module. This status data includes the current values of all sensors in the robot, as well as all the variables in the program. The controller module can also send commands to the monitor, which will be executed as soon as they arrive. These commands allow the controller to fully control the virtual machine, including directly manipulating all variables and pins, debugging the current program, or executing a new one.

The virtual machine implements a stack-based bytecode interpreter to execute the user programs. This implementation was chosen mainly because of its simplicity. Since the purpose of this project is educational, performance is not currently considered a high priority.

The GPIO module allows the user program and the virtual machine to access the pins, sensors, and actuators with a generic interface that facilitates porting the firmware to other microcontrollers, if necessary.

Regarding the programming tools, Mendieta uses a fork of Physical Bits,[2] a web-based programming environment for educational robotics that supports live coding and autonomy using a hybrid blocks/text programming language (Fig. 3).

Several changes to Physical Bits were needed for this project, the biggest being porting the middleware code from Squeak Smalltalk to Clojure. Our early tests showed that performance on the selected microprocessors (at the time, Raspberry Pi Zero W and Orange Pi Zero) was less than ideal and well below our requirements. We experimented

[2] Physical Bits website: https://gira.github.io/PhysicalBits/.

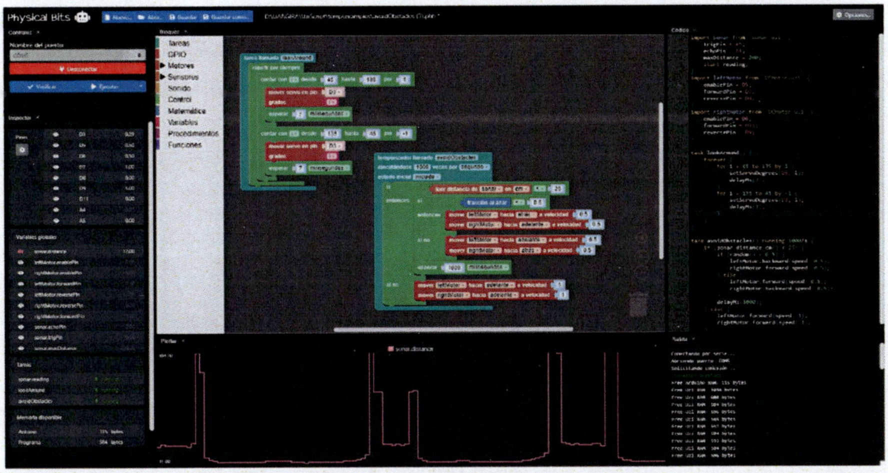

Fig. 3. User interface for robot programming

with different languages and eventually settled on Clojure due to its performance and language characteristics. In particular, the possibility of deploying the code on top of either Java or Javascript (thanks to ClojureScript), and the REPL and its live coding facilities (which reminded us of Smalltalk and its live image).

Additionally, to make a more efficient use of the microprocessor limited resources we are working on moving most of the compiler from the middleware to the clients. This way the cost of compiling the students' programs will be shared by all the computers in the classroom instead of falling into the microprocessor. For this change we are taking advantage of the Clojure and ClojureScript compatibility mentioned above.

We plan on making some other smaller changes that mostly involve changing the firmware to reduce the reporting frequency of pins and globals, as well as adding support for optical encoders and car movement primitives.

Considering that users are expected to spend most of their time working on their programs without immediate access to an actual robot (until their turn arrives), we are developing a simple simulator that would allow students to perform some basic validations for their programs while they wait for their turn to test in the actual robot. Although limited, we believe this to be an effective development technique and we have used it extensively for the development of the Physical Bits firmware.

In order to allow students to visualize both the sensors available in the robot and the variables defined in the program, a monitoring tool will allow the students to choose what variable they want to observe and display its value over time using different visualizations. This monitoring tool would be available to all students regardless of their respective turns, allowing students to monitor the robot values even while it is executing another student's program. An early prototype of the monitoring interface can be seen below:

In the case of complex programs, some kind of debugging tool is essential to allow stopping the execution of the program, executing each instruction step by step, and observing the status of the program in detail in order to understand and resolve errors in its logic (Fig. 4).

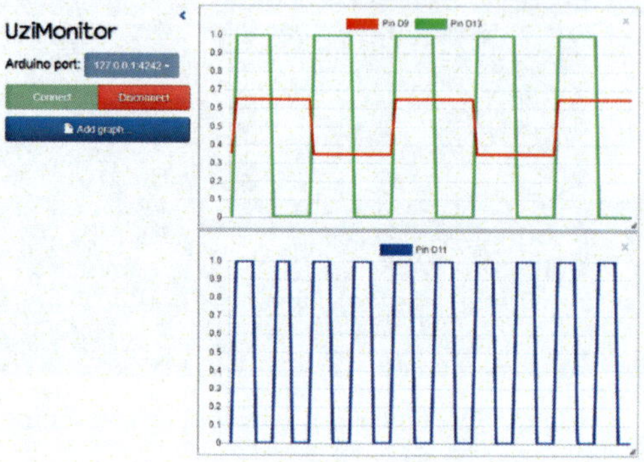

Fig. 4. User interface for monitoring variables and pins

This debugger prototype was fully functional in the Smalltalk version of the Physical Bits IDE but is not yet ported to Clojure (Fig. 5).

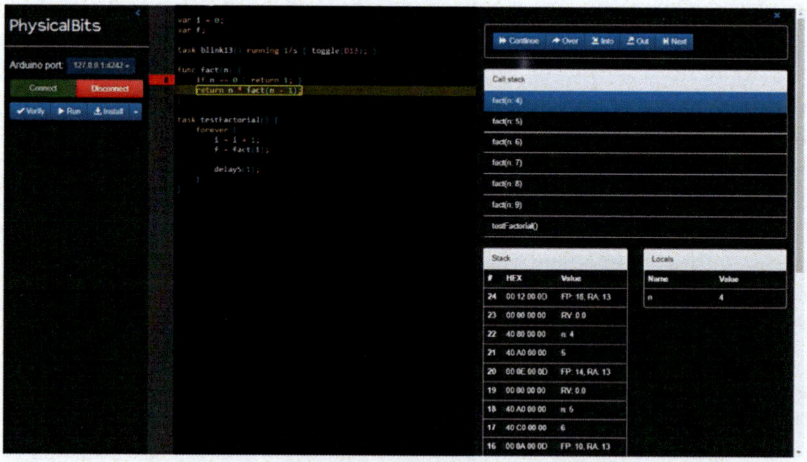

Fig. 5. User interface for debugging (early prototype)

Finally, the teacher will be able to access an administration interface that will allow him to view and control the queue of programs to be executed in the robot. This interface will allow to open and inspect the programs that were submitted to the robot, change their ordering, or even remove them from the queue. A prototype of the administration interface is shown below (Fig. 6).

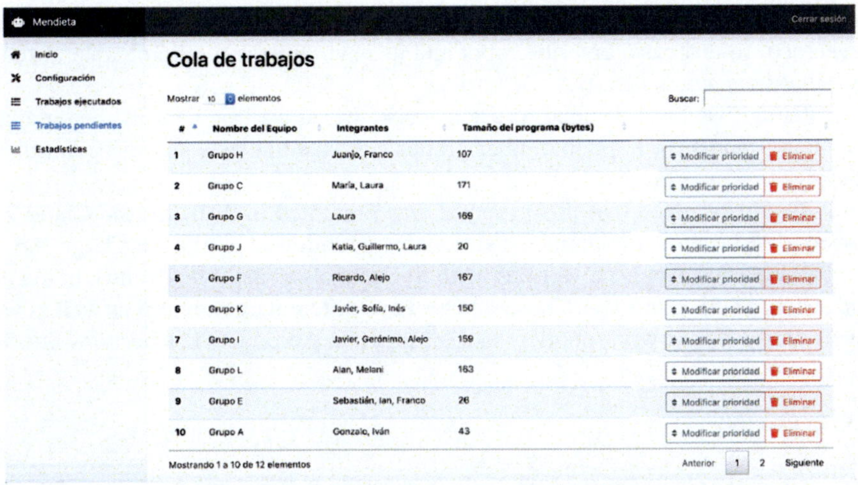

Fig. 6. User interface for managing the work queue

4 Schools' Involvement in the Pedagogical Decisions

Based on the principles of design thinking, for requirements analysis and testing cycles, we are working with five schools in our country with different characteristics. Some of them have experience using robots in the classroom, while the others are doing their first experience. In this way, in the first rounds of discussion, the view provided by the most experienced was essential. But at the same time, when we begin the testing cycles, inexperienced teachers will give us a perspective that will not be influenced by all previous experience and knowledge. In this way, by having the two groups of experimentation, we will be able to determine more clearly if the success or failure is given by our proposal or by the teaching experience.

Before designing the robot, we had four meetings with the schools to define the requirements for both hardware and software, which are what defined the characteristics previously presented. Subsequently, the emergence of the pandemic and the quarantine imposed in our country prevented us from carrying out the construction and testing stages, which we hope to be able to carry out this year.

5 Conclusions and Future Work

Mendieta is still in active development, we are currently working on a first prototype that will allow us to evaluate its effectiveness in the classroom. It is very likely that there will be changes to both its hardware and software architectures. Even though in this first iteration both the physical structure as well as the electronics components are well defined, the prototype we have developed so far is only a test model. More important than just this prototype is the framework we have defined that will allow building different robots under the same scheme of use (one robot per class).

It should be noted that it is possible to extend the use of Mendieta to support virtual education, by developing a site where students can watch the robot through one or more cameras and remotely program it using the same web tools that we have already developed. In that case, both the robot and its environment should be able to return to its initial state automatically.

Although we have decided to use a blocks-based environment as the default programming model we hope that other groups can take this as a starting point to create other types of interfaces.

It should be noted that this proposal was presented in 2019 for the Clarín Zurich Award for Education, obtaining a mention that has allowed to finance a large part of the project. We have also been working with five schools around the country in the design and testing of this prototype. These schools have different orientations as well as various levels of experience with robotics, so their input on the project has been invaluable.

References

Arduino Nano | Arduino Official Store. (s/f). Recuperado el 15 de junio de 2020, de. https://store.arduino.cc/usa/arduino-nano

Aroca, R.V., Gomes, R.B., Tavares, D.M., Souza, A.A.S., Burlamaqui, A.M., Caurin, G.A., Goncalves, L.M.: Increasing students' interest with low-cost Cell Bots. IEEE Trans. Educ. **56**(1), 3–8 (2012)

Banchoff Tzancoff, C.M., Martin, S., Gómez, S., López, F.E.M.: Experiencias en robótica educativa: Diez años trabajando con escuelas argentinas. XIV Congreso Nacional de Tecnología en Educación y Educación en Tecnología (TE&ET 2019), (Universidad Nacional de San Luis, 1 y 2 de julio de 2019) (2019)

Darrah, T., Hutchins, N., Biswas, G.: Design and development of a low-cost open-source robotics education platform. In: ISR 2018; 50th International Symposium on Robotics, pp. 1–4 (2018)

Erdemir, G., Kuzucuoglu, A.E., Kaplanoglu, E., El-Kahlout, Y.: Design and Implementation of Web Based Mobile Robot Control Platform for Robotics Education. Applied Mechanics and Materials; Trans Tech Publications Ltd. (2015). https://doi.org/10.4028/www.scientific.net/AMM.704.283

Finocchiaro, A.: Núcleos de Aprendizaje Prioritarios para Educación Digital, Programación y Robótica-Resolución CFE 343/2018

Freaks, E.: Ultrasonic ranging module hc-sr04. HC-SR04 datasheet (2016)

Handson Technology. (s/f). L298N Motor Driver. Recuperado el 15 de junio de 2020, de. http://www.handsontec.com/dataspecs/L298N%20Motor%20Driver.pdf

Hitachi. (s/f). HD44780U (LCD-II), (Dot Matrix Liquid Crystal Display Controller/Driver). 60

Hostler, A., Benson, B., Warner, J.: BudgetROV: an ultra low cost robotics platform for education and research. In: OCEANS 2017-Aberdeen, pp. 1–4 (2017)

Junior, L.A., Neto, O.T., Hernandez, M.F., Martins, P.S., Roger, L.L., Guerra, F.A.: A low-cost and simple arduino-based educational robotics kit. Cyber J. Multidisciplinary J. Sci. Technol. J. Sel. Areas Robot. Control (JSRC), December edition, **3**(12), 1–7 (2013)

Lopes Filho, J.A.B., Almeida, W.R.M., Martins, S.G.: Development of a multitasking mobile robot for the construction of educational robotics kits. In: 2011 International Conference on Electronic Devices, Systems and Applications (ICEDSA), pp. 213–216 (2011)

López-Rodríguez, F.M., Cuesta, F.: Andruino-A1: low-cost educational mobile robot based on android and arduino. J. Intell. Robot. Syst. **81**(1), 63–76 (2016). https://doi.org/10.1007/s10846-015-0227-x

MOCH22A Interrupter Datasheet pdf—OPTO Interrupter. Equivalent, Catalog. (s/f). Recuperado el 15 de junio de 2020, de. https://datasheetspdf.com/pdf/685501/Junye/MOCH22A/4

Motor Yellow 3-12VDC 2 Flats Shaft. (s/f). Wiltronics. Recuperado el 15 de junio de 2020, de. https://www.wiltronics.com.au/product/10137/yellow-motor-3-12vdc-2-flats-shaft/

Ministerio de Educación - Presidencia de la Nación. (2017). Escuelas del Futuro. Recuperado el 12 de junio de 2020, de. http://www.bnm.me.gov.ar/giga1/documentos/EL005852.pdf

Munizaga, R.A., Moleker, C., Yonzo, G., Zabala, G.: Everyone to robotics: an educational robotics project in Argentina. In: Robocup 2013 (2013)

NanJing Top Power ASIC Corp. (s/f). TP4056 1A Standalone Linear Li-lon Battery Charger with Thermal Regulation in SOP-8. Recuperado el 15 de junio de 2020, de https://www.semtronics.net/shop/resources/pdf_document/18/c1/9.pdf

orange pi zero—Orangepi. (s/f). Recuperado el 15 de junio de 2020, de http://www.orangepi.org/orangepizero/

Ripani, María Florencia. Programación y robótica: Objetivos de aprendizaje para la educación obligatoria. Ministerio de Educación - Presidencia de la Nación (2017). https://www.argentina.gob.ar/sites/default/files/programacion_y_robotica_0.pdf

Rubenstein, M., Ahler, C., Nagpal, R.: Kilobot: a low cost scalable robot system for collective behaviors. In: 2012 IEEE International Conference on Robotics and Automation, pp. 3293–3298 (2012)

Saleiro, M., Carmo, B., Rodrigues, J.M., du Buf, J.H.: A low-cost classroom-oriented educational robotics system. In: International Conference on Social Robotics, pp. 74–83 (2013)

Semiconductors, V.: TCRT5000, TCRT5000L, August 2009

Children's Debugging Processes and Strategies with a Simulated Robot: A Case Study

Dimitrios Nikolos(✉), Anastasia Misirli, and Vassilis Komis

Department of Educational Sciences and Early Childhood Education, University of Patras, Patras, Greece
{dnikolos,amisirli,komis}@upatras.gr

Abstract. A new simulated robot environment that focuses on the debugging process is used in this study. The environment provides debugging features such as fast execution of commands, the ability to delete single commands and trace. Two 7-year-old children used the environment, and their activities were recorded. Children had to move a simulated robot through a maze using move and direction Logo-like commands. The children used the debugging features of the environment efficiently and solved the problems assigned to them. They followed the debugging strategies such as identification of the bug, pinpointing the problematic code segment and correcting the program. The instructor was able to help them through the process when it was needed.

Keywords: Programming · Debugging · Simulated robots

1 Introduction

In this study, we are exploring a new Logo-like programming environment that is based on a simulated robot. This programming environment emphasizes the debugging strategies early primary school children use when they program.

Children need to learn about Computational Thinking (CT) and programming since these subjects have a significant role in many educational programs globally. Programming forms the modern world and cannot be neglected [1], and through programming and the development of CT children can adopt a new way of thinking, that can help them as a horizontal skill [2]. Computational Thinking incorporates debugging along with other dimensions such as: abstraction, algorithmic thinking, automation, decomposition, and generalization [3]. In this study we used a simulated programmable roaming robot [4] as a vehicle for investigating debugging strategies.

Bugs do exist in environments intended for young children [5]. Debugging is the process of identifying why a program misbehaves and correcting it [6]. It is an essential part of every programming activity, and educators view it as an inseparable part of the process of understanding a program [7]. The debugging strategies the children incorporate when they correct errors are of interest, especially since the programming environments are changing through the years, with the biggest change being the advance of block-based programming [6].

M. Malvezzi et al. (Eds.): EDUROBOTICS 2021, SCI 982, pp. 64–74, 2021.
https://doi.org/10.1007/978-3-030-77022-8_6

Scaffolding is the main strategy to support and develop the debugging skills of young children. Scaffolding can be provided through visualization [8] or by the teacher [9]. The environment of this study is designed to provide visualization of the programming process so that the teacher can intervene when it is needed and is a simulated roaming robot.

Physical roaming robots are used for introducing Computational Thinking to pre-school and first grades of primary school children [10]. Such robots include the Bee-Bot, Pro-Bot, and Finch. Some of these robots have simulated counterparts, e.g. the Bee-Bot app for Android and iOS. The first Logo turtles were roaming robots and their on-screen counterparts reflect robot behavior [11].

A simulated robot for education is an old concept, Karel the programming language and robot was created in the 80s [12] without a physical counterpart. After many years, a modified version that accommodated modern concepts was still in use [13]. As the growth of online learning is accelerated by the COVID19 pandemic, we expect simulated robots to gain a place on CT development through remote means. However, our approach did not stem from the need for an online counterpart of real robots, but from the need to visualize the algorithm of the robot and the problem-solving strategies the children demonstrate while they program. In other words, we were driven from the need to facilitate and support children in constructing abstraction and decomposition skills through visual means.

Our simulated robot [4] visualizes its program that is hidden in most physical roaming robots, for the children to be able to edit it in a visual way. The editing process is visible to the teacher so that he may be able to assist them if needed.

A curriculum that is used for the development of CT and incorporates roaming robots and the Scratch programming language [14] is the context of this study. To facilitate evaluation of the concepts that we teach, a version of a simulated robot that follows the paradigm of a Logo-like environment was created [4]. This simulated robot environment provides a set of features that facilitate debugging namely, a) unique starting state, b) fast execution c) deletion of a single command, and d) trace.

The following research questions guided our case study:

1. Do young children use the debugging features of the environment when they are programming using it?
2. What strategies do children incorporate during debugging?
3. Do children develop decomposition skills during debugging?

In the next section the methodology is described, in the third section the findings are drawn, discussion and conclusions follow.

2 Methodology

To explore the possibility of using the simulated robot environment as a tool for debugging activities a case study with two seven-year-old children is conducted. The actions of the children were being recorded and the recordings compose the research data. Informed and signed consent forms were obtained from parents and guardians.

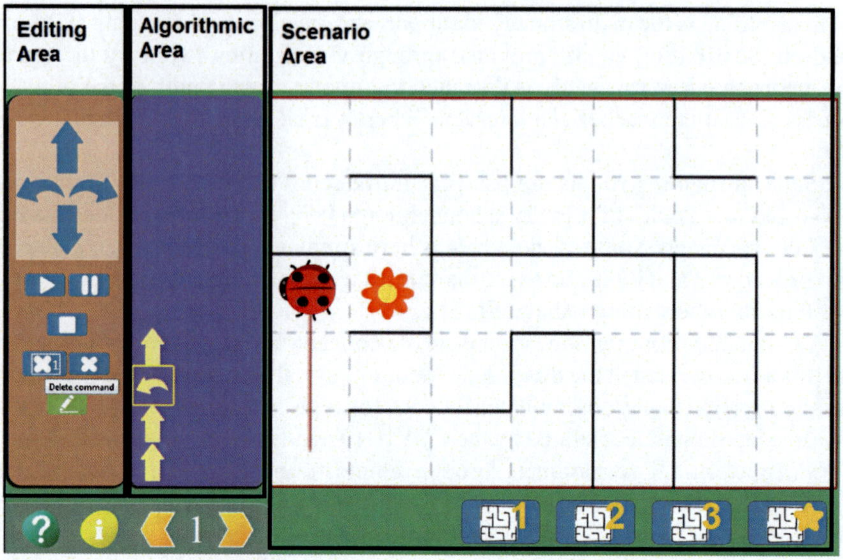

Fig 1. The digital environment with the three areas

2.1 The Environment

The digital environment that was used consists of three areas: a) editing area, b) algorithmic area and c) scenario area (Fig. 1).

In [8] two scaffolding strategies are studied for debugging with the Bee-Bot. The first strategy involves cards with the Bee-Bot commands an approach like the one found in [14, 15]. The second strategy involves writing down commands on a piece of paper. The environment of this study provides a visual representation that resembles the visual representations that both these strategies produce [4].

The environment follows the paradigm of Scratch 3 where the available commands are positioned on the left, the program is formed in the middle and the results are shown on the right.

The editing area features the available commands, and the controls. The available commands are Forward (FW), Backward (BW), Left (LT) and Right (RT). These commands move the simulated robot (ladybird) according to its direction. The ladybird turns in place following the Logo philosophy.

The controls include (Fig. 1):

1. A Play button that executes the program,
2. A Pause button that pauses the execution of the program,
3. A Stop button that stops the program and returns the ladybird to its (default) starting point and direction (lower left cell, looking up)
4. A ×1 button that deletes only the selected command,
5. A × button that deletes the program,
6. A pencil button that toggles the trace the ladybird leaves behind (if it is active the ladybird leaves trace, otherwise the ladybird does not leave a trace).

The features that facilitate the debugging process are shown in Fig. 1. The unique starting point means that if the Play button is pressed the ladybird will return to the first cell. The fast execution is also shown, when a command is selected the program runs up until that command, and the ladybird is transferred to the computed cell. Deletion of a single command is supported. In the example of Fig. 1 one possible action would be to delete only the undesired selected command with the $\times 1$ button and replace it with the correct one. In Fig. 1 the trace is enabled (pen down in Logo).

2.2 Sample

Our sample for this study is two children a boy and a girl that were 7 years old. The girl used the environment on a tablet, while the boy used it on a computer. Both children were familiar with the devices they were using since they used them at their home. They were not familiar with any digital programming environment.

3 Findings

3.1 The Nature of the Activity

In all cases, both the girl and the boy identified the path the ladybird should follow and pointed with their finger accordingly. The problem the children had to solve is the translation of the path to commands. In our context the decomposition of the problem into smaller ones was done by the children. In many cases they focused on the next step the ladybird had to perform. The general problem that was to translate the path to commands was decomposed in many next-cell problems.

We encoded the next-cell problems with the encoding depicted in Table 1. The first letter represents the direction of the ladybird and the second letter represents the direction of the cell it needs to go to. When the ladybird is looking Up (U) and must go to the cell that is next to it to the right (R), we have a UR problem. When the ladybird is looking down and must go down, we have a DD problem.

For the UU, RR, DD and LL problems the solution is a FW command, e.g. if the ladybird is looking to the right and wants to go to the cell that is next to it to the right, it has to move towards its own direction. For the RU, DR, LD and UL problems the solution is an LT command followed by a FW command, e.g. if the ladybird is looking up and wants to go to the cell that is next to it to the left, it must turn to its left and then move forward to its new direction. In a similar manner, for the UR, RD, DL and LU problems the solution is an RT command followed by an FW command. Finally, for the UD, DU, LR and RL problems a solution could be a BW command since the ladybird is looking on the opposite direction of where it must go.

3.2 Children's Processes and Strategies

The girl pointed with her finger through the correct path for the solution of the maze of Fig. 2, trying to organise the correct algorithm. In the actual activity the numbers were not shown in Fig. 2, and the girl was viewing the maze without these numbers. The goal

Table 1. Next-cell problems

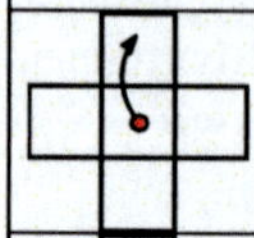	UU	RU	DU	LU
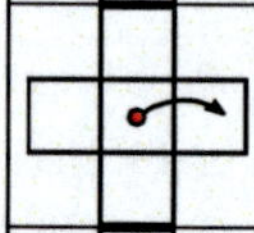	UR	RR	DR	LR
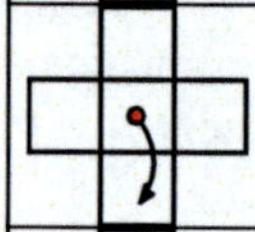	UD	RD	DD	LD
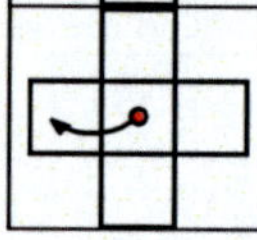	UL	RL	DL	LL

of the task was to write a program that makes the ladybird follow the path. The cell numbers were added for our analysis.

The debugging episodes we identified are summarized in Table 2. Each problem is characterized by its target-cell and the problem that had to be solved to get there.

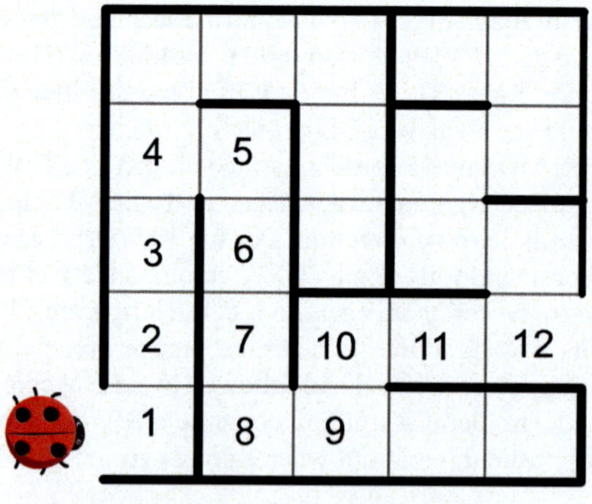

Fig. 2. The maze the girl had to solve

In the first row of Table 2 we see that the girl tried to move the ladybird to the right using the RT command. She also tried to move it to the next steps using three FW commands. The instructor helped her to execute the commands and to delete the single

Table 2. Debugging episodes (girl)

Target Cell	Next-cell problem	Optimal solution	Girl's first try	Girl's final solution	Environment features used
1	*RR*	FW	RT, FW, FW, FW	FW	Delete one command, fast execution
2	*RU*	LT, FW	FW, FW	LT, FW	Delete one command, fast execution
6	*RD*	RT, FW	LT	LT, RT, RT, FW	Fast execution, play button
10	*RU*	LT, FW	RT, LT, FW	LT, FW	Delete one command, fast execution, play button
11	*UR*	RT, FW	FW	RT,FW	Delete one command, fast execution, play button

command that causes the problem using the x1 control button. The girl decomposed the actions needed to move from its current position to cell 4 but was not able to perform the first step because she added the RT command.

In the second row of Table 2, the girl had to solve the RU problem, but because the ladybird must go up, she used the forward command. The instructor showed her where the new command is inserted when a command is selected. The problem at hand was to move from the current position to the next cell, the cell above in particular. In that context, the commands that she left in the program after the necessary for the RU problem seem irrelevant. She deleted all the trailing commands later.

In the third row of Table 2, the girl is not able to select the correct turn command. When she realizes that the command does not provide the desired outcome, she adds two of the correct turns, the first one neutralizes the error and the second one is correcting the program. She has left these two extra commands through the end of the activity. When asked about them she justified the moves of the ladybird by stating "it's like I had left something for it to eat".

The most challenging task for the girl was the RU problem she had to solve when she was trying to go to the 10th cell, and it is shown in the fourth row of Table 2. Even though she had solved the same problem in previous steps, she found it very difficult to solve it in this context. The main reason for this was that she had to find the problematic areas of the program while the program was already large for her (around 15 commands) and the fact that she added commands that were to be executed after the 10th cell. Another factor that made this problem difficult for the girl was that she confused left and right very often especially when the ladybird was not looking up.

In the last line of Table 2 the girl had to solve the UR problem and she did that without help from the instructor. After the girl the successful solution of the problem the instructor prompted her to delete the extra FW commands.

Fast execution played an important in all the debugging episodes. The girl often moved in different commands of her program to test the effect. The delete one button ($\times 1$) was the main tool that was used for debugging, but not always. In one case she debugged the program by adding more commands.

In most steps the girl was confident that her program should work, and it came as a surprise when it did not.

The boy followed a trial and error strategy and even when he was able to articulate a solution he did not form the (correct) program as a whole but he decomposed it to one or two commands and run them to see the result. An example of the incremental way of thinking the boy applied is the following: The boy had to form the program RT, FW, LT, FW, FW. He formed the program RT, FW, LT, FW (the last FW was missing). The dialogue followed:

Instructor: *Are these commands enough?*

Boy: *Let me see.* [hits the run button].

Later in the activity he run the program in various occasions and said: "Let me see this one and then I will know what to do" and then again "I will do the thing I did before, I see what will happen and then I will continue". This last sentence indicates that he uses a trial-and-error strategy.

Before learning how to delete a single command he deleted the whole program and retried several times. In one case he deleted the whole program, rewrote the exact same program and was very disappointed to see the same result. After the instructor showed him that he can delete one command he immediately used it to correct another error of the program. But when he was left alone to correct a program, he deleted the last commands until the point of the problem and then inserted the commands he wanted.

As was the case with the girl the boy also had problems with the turn commands. In one case he said that no matter what turn command he would choose the effect will be the same. Only after the intervention of the instructor he seemed to understand that the commands would have different effect. In another point the boy used two extra correct turn commands to correct a turn with the first neutralizing the error and the second correcting the program, the girl had done the same.

The strategies they incorporated were different; the girl solved the problems by moving to neighbor cells while the boy tried to solve larger problems involving the cells after the neighbor one. The boy used a trial and error strategy while the girl tried to find the correct solution to every problem. Both the girl and the boy had problems with the left and right direction commands and they both solved a problem by adding two correct direction commands instead of replacing the wrong one.

4 Discussion

As far as the appropriateness of the environment is concerned, we think that the environment facilitated debugging and provided a tool that surfaced the debugging strategies the children used. They had available tools to a) identify the position of the bug (fast execution), b) locating the bug and acknowledge there is a problematic situation (fast execution), c) fix the problem (delete one command, insert one or more commands after the selection), and they actually used them purposefully.

The pause button was not used neither by the children nor by the instructors in this study. The stop button was used as a restart button, a metaphor that is also used in media players. The delete all button is also useful because in some cases children had to delete the whole program, as it would have been more difficult to fix it by deleting single commands.

As far as the debugging strategies are concerned, the children tried to solve the problem, viewed the result, and acted accordingly. This behavior is supported by the environment, but it is also the way children tend to operate in programming environments [16, 17]. In [18] some of the steps that are involved in the debugging process are presented. In our debug-oriented environment these steps are adjusted as follows:

1. Obtain and understand the meaning of the symptoms of the buggy program: in our environment such a symptom would be that the simulated robot does not reach the target or does not move to the desired direction.
2. Identify the appropriate code segment(s) involved in the bug: in our environment the child can click on a command, see the position it corresponds to (fast-execution) and identify the code that creates the problem
3. Identify the process in which the bug occurs: in our environment the child creates only one program (one process). Therefore, this step does not apply to our analysis.
4. Know the most likely cause of the bug: in our environment this is equivalent to finding the position of the bug (step 2) and does not apply to our analysis.
5. Evaluate the evidence and select a strategy for correcting the error: in our environment the child can either delete the whole program, delete the wrong command to replace it or add more commands.

Only the first, the second and the fifth steps are necessary in our study, since there is only one program (one process) and the cause of the bug is wrong commands on the position they found in step 2. These are like the steps that are described in the meta-analysis of [19] as the cycle observe, hypothesize, modify, and test.

We have found that our children follow these steps as they debug. For the first step they might not articulate that they found a symptom of a buggy program with appropriate vocabulary, due to their age. They usually used an exclamation such as 'uh-oh', 'not again', 'the same happened again' or other expressions of disappointment.

The second step was the most difficult for the children. The fact that the programs were simple does not mean that young children can pinpoint problematic segments easily [20]. During the children's first programming efforts they experience difficulties in isolating the problematic segments since they thought that the following commands were erroneous also. This observation is in accordance with [16], the children of their research thought that "all blocks after the bug were misbehaving."

Finally, for the last step our children used three strategies for correcting the bug:

1. A strategy that was followed early on the activity was the deletion of the whole program (start over). This strategy was not followed later mostly because they did not want to lose the work they had already done.
2. In the second strategy they deleted the wrong command and replaced it with the one they wanted. This strategy was developed after instructor intervention for the function of the x1 control button.
3. The third strategy was to add commands to correct the behavior of the wrong commands in accordance with the findings of [21].

More research is needed for program visualization to be proved the key for program debugging [22], but for early years and first grades of primary school children, visualization is necessary as a scaffolding technique. Brusilovsky suggested the following human intervention for debugging [23]: a) demonstrate an instance in which the code fails, b) provide a visual execution of the code, c) explain the code while providing a visual execution, and d) provide additional help. This is the scaffolding technique the instructors used with our environment. It is important to mention that the children created the instance in which the code fails. The instructor helped them through the debugging process when needed. In many cases, they were able to debug by themselves.

As far as decomposition is concerned, we noticed that the girl decomposed the problem in small segments (moving to the next cell) whereas the boy tried to solve bigger problems (i.e. solving the next turn and following forwards). The decomposition strategies of both the girl and the boy were bottom-up. They first moved towards the correct direction since they had already pointed the correct path with their finger. After that, they tried to build the path from the position they were found at the time. When there are more commands in the program, especially if they are after the bug, they found it more difficult to debug.

5 Limitations

The small sample is the main limitation of this study. Because of the pandemic, children and parents are having difficulties in organizing the factors that are involved in the learning process. Another implication is that every child was alone with the teacher when they were using the environment. The communication between children when they debug is an important factor [24]. For evaluating debugging competencies having a more elaborated design of cognitive analysis and assessment, along with a bigger sample would be appropriate [18, 20], ideally in the classroom setting.

6 Conclusions

The environment of [4] seemed appropriate for surfacing the debugging activities of children and providing context for teachers to help with these strategies. Our children followed the debugging strategies that are found in the bibliography such as identification of the bug, pinpointing the problematic code segment and correcting the program.

The visualization features of the environment helped the instructors apply appropriate scaffolding techniques for reinforcing the debugging skills of the children. Finally, the children decomposed the problems in a bottom-up manner.

A feature that would help would be a way to visualize the correct path that is found by the child, but it is not shown on screen. This feature will help children to not have to recall the path from their limited memory resources [8] and will help the educator to distinct between errors in path and errors in command. Though it was easy for our children to identify the path, when the program becomes large it would be useful to have a visual aid of the direction they have to go to.

Other debugging strategies might emerge if we had ready-made programs for our children. With ready-made programs, the children would focus on the debugging processes [19] rather than creating the program that creates cognitive load. Children would organize their thought on the difference between the intended outcome and the actual outcome [19] this thinking process is a key difference between experts and novices [18]. Activities with ready-made programs could be included in our curriculum at a later stage.

The environment of [4] seemed appropriate for studying the debugging strategies of young primary school children and more research is needed to explore the different debugging processes and strategies the children may apply when they program the simulated robot.

Acknowledgement. This research is co-financed by Greece and the European Union (European Social Fund ESF) through the Operational Programme 'Human Resources Development, Education and Lifelong Learning 2014–2020' in the context of the project 'Robotics and programming languages: how to develop Computational Thinking in Preschool and Primary Education' (MIS 5047126).

References

1. Rushkoff, D.: Program or be Programmed. Or Books, New York (2010)
2. Wing, J.: Computational thinking. Commun. ACM **49**(3), 33–35 (2006)
3. Bocconi, S., Chioccariello, A., Dettori, G., Ferrari, A., Engelhardt, K.: Developing computational thinking in compulsory education: Implications for policy and practice. Resource document, EUR 28295EN (2016)
4. Misirli, A., Nikolos, D., Komis, V.: Transferring experiences in logo-like environment in computational thinking game design. In: Proceedings of the 13th Annual International Conference of Education, Research and Innovation, pp. 9721–9728 (2020)
5. duBoulay, B.: Some Difficulties of learning to program. J. Educ. Comput. Res. **2**(1), 57–73 (1986)
6. Kim, C., Yuan, J., Vasconcelos, L., Shin, M., Hill, R.B.: Debugging during block-based programming. Instr. Sci. **46**(5), 767–787 (2018)
7. Papert, S.: Mindstorms: Children, Computers and Powerful Ideas. Basic books, New York (1993)
8. Angeli, C., Valanides, N.: Developing young children's computational thinking with educational robotics: an interaction effect between gender and scaffolding strategy. Comput. Hum. Behav. **105**, 105954 (2020)

9. Wang, C., Choi, Y., Benson, K., Eggleston, C., Weber, D.: Teacher's role in fostering preschoolers' computational thinking: an exploratory case study. Early Educ. Dev. **32**(1), 26–48 (2021)
10. Misirli, A., Komis, V.: Robotics and programming concepts in early childhood education: a conceptual framework for designing educational scenarios. In: Karagiannidis, C., Politis, P., Karasavvidis, I. (eds.) Research on e-Learning and ICT in Education. Springer, New York (2014)
11. Solomon, C., Harvey, B., Kahn, K., Lieberman, H., Miller, M., Minksy, M., Papert, A., Silverman, B.: History of logo. In: Proceedings of the ACM on Programming Languages, 4(HOPL) (2020)
12. Pattis, R.E.: Karel the Robot: A Gentle Introduction to the Art of Programming. Wiley, Hoboken (1981)
13. Becker, B.W.: Teaching CS1 with karel the robot in Java. In: Proceedings of the 32nd SIGCSE Technical Symposium on Computer Science Education, SIGCSE, pp. 50–54. ACM, New York (2001)
14. Komis, V., Romero, M., Misirli, A.: A scenario-based approach for designing educational robotics activities for co-creative problem solving. In: International Conference EduRobotics, pp. 158–169. Springer, Cham (2016)
15. Misirli, A., Komis, V.: Emerged debugging abilities in early childhood education. In: The Proceedings of Constructionism Moving Forward (2020)
16. Sipitakiat, A., Nusen, N.: Robo-blocks: designing debugging abilities in a tangible programming system for early primary school children. In: Proceedings of the 11th International Conference on Interaction Design and Children, pp. 98–105 (2012)
17. Fessakis, G., Gouli, E., Mavroudi, E.: Problem solving by 5–6 years old kindergarten children in a computer programming environment: a case study. Comput. Educ. **63**, 87–97 (2013)
18. Xu, S., Rajlich, V.: Cognitive process during program debugging. In: Proceedings of the 3rd IEEE International Conference on Cognitive Informatics, ICCI 2004, pp. 176–182 (2004)
19. Rich, K.M., Strickland, C., Binkowski, T.A., Franklin, D.: A K-8 debugging learning trajectory derived from research literature. In: Proceedings of the 50th ACM Technical Symposium on Computer Science Education, pp. 745–751 (2019)
20. Cuneo, D.O.: Young children and turtle graphics programming: generating and debugging simple turtle programs. In: Annual Meeting of the American Educational Research Association (1986)
21. Murphy, L., Lewandowski, G., McCauley, R., Simon, B., Thomas, L., Zander, C.: Debugging: the good, the bad, and the quirky-a qualitative analysis of novices' strategies. ACM SIGCSE Bull. **40**(1), 163–167 (2008)
22. McCauley, R., Fitzgerald, F., Lewandowski, G., Murphy, L., Simon, B., Thomas, L., Zander, C.: Debugging: a review of the literature from an educational perspective. Comput. Sci. Educ. **18**(2), 67–92 (2008)
23. Brusilovsky, P.: Program visualization as a debugging tool for novices. In: Ashlund, S., Mullet, K., Henderson, A., Hollnagel, E., White, T. (eds.) Proceedings of INTERACT 1993 and CHI 1993 Conference Companion on Human Factors in Computing Systems, pp. 29–30. ACM Press, New York (1993)
24. Heikkilä, M., Mannila, L.: Debugging in programming as a multimodal practice in early childhood education settings. Multimodal Technol. Interact. **2**, 42 (2018). MDPI

AI-Robotics and AI Literacy

Amy Eguchi(✉)

University of California, La Jolla, San Diego, CA 92093-0070, USA
a2eguchi@ucsd.edu

Abstract. AI is considered as a rapidly advancing technological domain capable of altering every aspect of our lives and society. However, many developing countries lack the basic infrastructure needed for taking advantage of AI systems, which could worsen the new technological, economic, and social divides. Ensuring the inclusion and equity of AI in education is one of the challenges that we need to solve. How can we support the urgent needs of developing countries and the underprivileged communities so that the divide between "haves" and "have-nots" will not continue to grow? This is the core question that this project is trying to address. This paper introduces the AI-robotics project focusing on developing an open-source affordable AI-robotics tool to address the need to promote AI literacy around the world. CogBots, the AI-empowered educational learning tool, has been in development collaboratively with Google, CogLabs and UNESCO. The project targets upper elementary and middle school students to promote AI literacy, including the 5 Big Ideas in AI, using the AI-powered educational robotics tool and the Teachable Machine by Google Creative Lab providing a machine learning experience.

Keywords: Artificial intelligence · AI-robotics · AI literacy

1 Introduction

In 2019, the UNESCO highlighted the issues the world is facing as follows:

- the world is facing a learning crisis (currently, more than 200 million children globally are out of school, while a further 600 million are in school but are not achieving minimum levels of competency);
- it is education that powers sustainable development (education is a key driver of all the UN Sustainable Development Goals (SDGs), as well as the focus of SDG 4); and
- stainable development will only be achieved if we successfully leverage the digital revolution, which now increasingly includes artificial intelligence (AI) [1].

AI is considered as a rapidly advancing technological domain capable of altering every aspect of our lives and society. AI has been influencing our lives where data sets are available and improvements by either automation or inventions are needed [2]. Many children in developed countries are now growing up with AI-assistants or AI-assisted smart devices in their homes (i.e. Google enhanced smart speaker, the devices equipped

M. Malvezzi et al. (Eds.): EDUROBOTICS 2021, SCI 982, pp. 75–85, 2021.
https://doi.org/10.1007/978-3-030-77022-8_7

with Siri or Alexa). The rapid development of voice and facial recognition technologies, and machine learning algorithms have already transformed our everyday lives.

Moreover, such AI technologies will continue to evolve with astonishing speed. It is evident that AI has become ubiquitous in our society including K-12 education. AI technology to enhance education by producing new learning and teaching solutions has been tested in the field [3]. AI has the potential to improve the education system by using data to promote equity and high-quality education in developing countries. However, to reach the goal, AI requires advanced infrastructures and an ecosystem that can produce creators and innovators who can contribute to the development of future AI systems. Many developing countries lack the basic infrastructure needed for taking advantage of AI systems, which could worsen the new technological, economic, and social divides. Ensuring the inclusion and equity of AI in education is one of the challenges that we need to solve. How can we support the urgent needs of developing countries and the underprivileged communities so that the divide between "haves" and "have-nots" will not continue to grow? This is the core question that this project is trying to address.

There is an emerging need for teaching AI in schools. Educators across fields from computer science, AI, to education from around the world are already addressing the urgent need to help people understand the science behind AI, its limits, and its potential societal impacts in our everyday lives as well as in the future. What is specifically crucial is to prepare K-12 students for AI-powered context and their future professions, many of which might not exist yet, and become citizens capable of understanding and utilizing AI-powered technologies to ensure the equity and inclusion of *all* people they serve [3, 4].

This paper introduces the AI-robotics project focusing on developing an open-source affordable AI-robotics tool to address the need to promote AI literacy around the world.

2 AI-Robotics Tool and AI Literacy

2.1 AI-Robotics Tool and AI Literacy

Literacy empowers people by enabling them to communicate with others, to express and understand various opinions and ideas, to help them make informed decisions and solve problems that they face, and to enable them to participate in society purposefully. Similarly, AI literacy empowers people to actively participate in society through informed decision making, while understanding AI's influences in society and its future potential, both negative and positive, as well as contributing to the development of an inclusive AI-ecosystem where all members of the society participate to better its future.

In the US, the AI4K12 initiative (https://ai4k12.com/), jointly sponsored by the Association for the Advancement of Artificial Intelligence (AAAI; https://aaai.org/) and the Computer Science Teachers Association (CSTA; https://www.csteachers.org/), was formed to develop (1) national guidelines for AI education for K - 12, (2) an online, curated resource directory to facilitate AI instruction, and (3) a community of resource and tool developers focused on the AI for K-12 population of students. AI4K12 introduced the 5 Big Ideas in AI [4] in 2019 (Fig. 1):

1. Perception: Computers perceive the world using sensors,
2. Representation & Reasoning: Agents maintain representations of the world and use them for reasoning,
3. Learning: Computers can learn from data,
4. Natural Interaction: Intelligent agents require many kinds of knowledge to interact naturally with humans, and
5. Societal Impact: AI can impact society in both positive and negative ways.

Although there are several AI-enhanced online tools that support students' learning of AI concepts, the concepts remain rather abstract for students since their work stays inside of the computer (virtual and abstract). Students, especially younger ones, tend to struggle to fully grasp abstract concepts and ideas without applying them in real-life situations using manipulatives.

Seymour Papert's constructionism theory as its foundation, the learning environment that robotics tools create promotes students' learning of abstract concepts and ideas [5–7]. Using AI-robotics tools as manipulatives, abstract concepts and ideas become visible while students apply the concepts in real-life situations. Since the subject and concepts that AI introduces are foreign to many, making AI and its concepts visible through AI-robotics tools could help their construction of new knowledge, make their understanding of AI deeper, and help them retain the newly constructed knowledge.

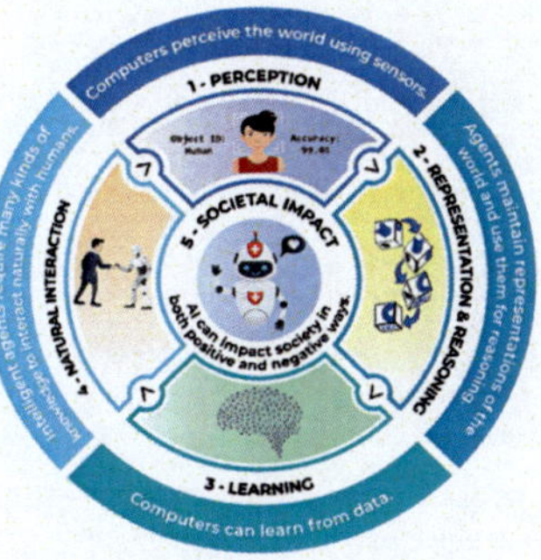

Fig. 1. 5 Big ideas in AI

3 AI-Robotics Tool

The AI-robotics tool in development aims at providing an affordable, open-source, AI-enhanced educational tool that can be accessible in any school in the world. It will be accompanied by sample lessons, curriculum ideas, and teaching materials to help teachers from around the world incorporating AI literacy into their lessons. Moreover, the tool will provide students hands-on machine learning experience with the manipulative.

The current version of the AI-robotics tools targets upper elementary to middle school students. CogBots, the AI-robotics tool in development, uses a model created with the Teachable Machine by Google Creative Lab. Using the image classification model, students can program CogBots using Scratch with ThinkBot extension (https://storage.googleapis.com/dev-thinkbot/index.html). Scratch Extensions allow Scratch projects to connect with external hardware, online tools (sources of information), or blocks with advanced functionality. The following sections provide more details of CogBots and how they work. CogBots can initiate inquiries and curiosities among students to explore AI technologies in everyday lives.

3.1 CogBots

Cogbots (Fig. 2 and 5) are open-source robotics kits developed by CogLabs in collaboration with Google and UNESCO. CogBots empower students through creating designs, assembling, and programming their own "thinking" robot. It uses sustainable and readily available materials like recycled boxes and cardboards, 3D printed parts, and recycled smartphones. Figure 3 shows the parts for the 3D printed robot model. If students have access to 3D printers, they can design the robot's body by tinkering the shape and/or changing the color of the parts to personalize their robots. In addition, it ignites students' creativity in their own robot design using a CAD app, such as Tinkercad (https://www.tinkercad.com/), and 3D print the parts. If students have no access to 3D printers or prefer working with craft materials, they can use cardboard or cardboard boxes with crayons, yarns, origami papers, paints, or papier-mache´ to create a robot's body (Fig. 4).

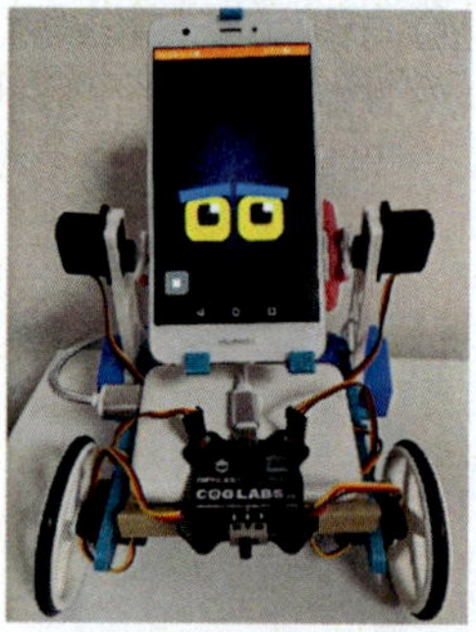

Fig. 2. CogBot

Fig. 3. CogBot parts

Fig. 4. CogBot with cardboard and craft materials

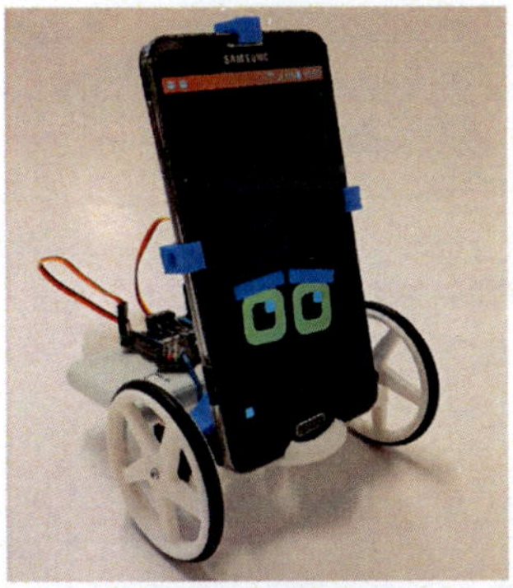

Fig. 5. CogMini

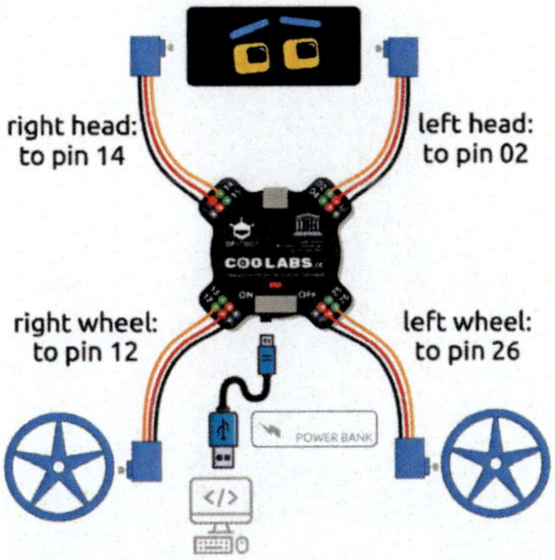

Fig. 6. CogMini connection diagram

Since the original CogBot (Fig. 1) requires students to use small screws and wire pins (for motor connectors), CogMini (Fig. 5) was developed for younger students or those who have less experience building with screws and breadboards. CogMini still requires small screws but fewer number of screws are needed. Also, the total number of parts it uses is also smaller. The building instructions (CogBot instruction: https://bit.ly/CogBot, CogMini building instruction: https://bit.ly/CogMini, Fig. 6 and 7) show how easy it is to put together the CogBots.

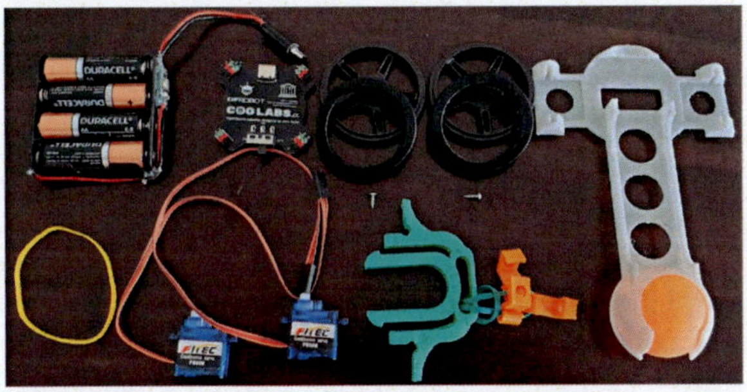

Fig. 7. CogMini parts

For motors, CogBot uses four micro 360-degree continuous rotation servo motors that are robust enough for agile movements with less budget. Two motors connected to wheels control the robot's movements, and the other two motors connected to a phone holder control the angle of the smartphone. CogMini Robot uses two micro servo motors that control the robot's movements. The price of the micro 360-degree continuous rotation servo motors is around US$5 a piece.

3.2 CogBot Controller

For the controller, CogBots uses an Arduino board that communicates with motors and a smartphone. Currently, there are two versions of Arduino boards that can control

Fig. 8. ESP32 boards

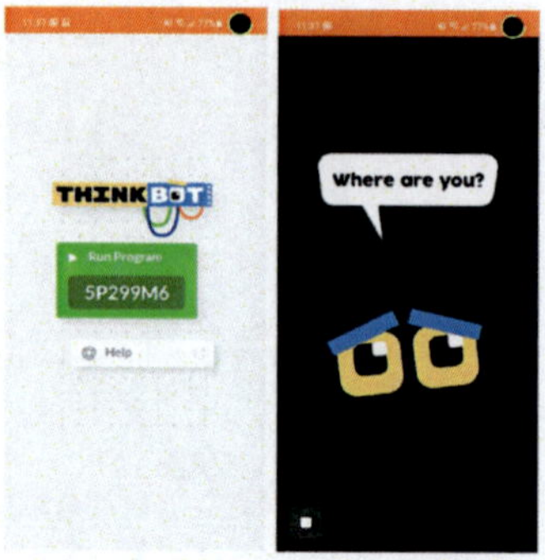

Fig. 9. ThinkBot application

CogBots - DOIT ESP32 DevKit v1 and CogLabs' ESP32 board (by DFRobot) (Fig. 7). They are ESP32 boards, a low-cost, low-power microcontroller board with integrated WiFi and dual-mode Bluetooth communication capabilities, which is robust enough for the AI-robotics tool. It helps to keep the overall cost lower, ideal for the wider and inclusive implementation of AI literacy education around the world. DOIT ESP32 DevKit v1 is less than US$10. DOIT ESP32 DevKit v1 uses a breadboard to connect motors, while CogLab's ESP 32 board comes with 3-pin connections to securely connect motors. CogMini uses CogLab's ESP 32 board targeting younger students, and can be built with fewer 3D printed parts and motors (Fig. 6).

3.3 CogBot Sensor

CogBot uses a smartphone as its sensors. The smartphone needs to be connected to the same local WiFi network that a computer for programming is connected to. A smartphone can be recycled one without a cellphone data subscription (Android 5.0 onwards). It communicates with a computer via a local WiFi network and with the ESP32 board via Bluetooth. The current version of CogBot and CogMini work only with Android phones using its camera. The plan is to utilize all other sensors that smartphones are equipped with, such as proximity sensor, light sensor, accelerometer, and thermometer.

3.4 Working with CogBots

A cellphone application, called ThinkBot, works as the robot's interface that can activate the code created on the computer. ThinkBot App works as the face of the robot showing facial expressions and statements as texts (Fig. 9). Currently, the ThinkBot App works only with Android phones. An iPhone application is under development and will be available in the future.

3.5 CogBots Coding Application

Scratch is recognized as one of the most powerful, popular, and accessible block-coding tools available in many countries around the world, which is an ideal application for CogBots. CogBots provides a machine learning experience through programming to control the AI-powered robot for students to learn AI concepts. Using the Scratch extension to make AI-CS experience accessible for students with little or no coding experience (Fig. 10). Scratch is beneficial because it is a web-based programming environment where no installation is required, making it easy for teachers to use in their classrooms since any installation on their classroom computers require IT support. Since Scratch is widely used in K-12 classrooms around the world, it will make programming with CogBots accessible to many students who have experience coding with Scratch.

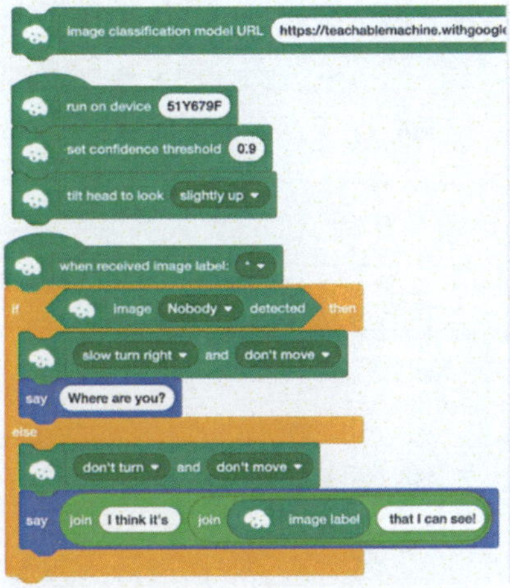

Fig. 10. ThinkBot scratch extension

The additional tool that enables CogBot to provide a machine learning experience is the Teachable Machine (https://teachablemachine.withgoogle.com/train/image) by Google Creative Lab. ThinkBot Scratch extension allows students to use an image classification model created with Teachable Machine in their codes to control the CogBots.

First, students capture image samples on the Teachable Machine website for training a model (Fig. 11). Then, students need to name each group of image samples (adding a label). Once students created more than one label with image samples, they train an image calcification model using the labeled image samples (Fig. 12). Next, students export the model. Once the model is exported, students can use the shareable URL with their Scratch code (Fig. 13).

To use the created image classification model, students paste the sharable link from the Teachable Machine website to the Image Classification Model URL block so that their code can retrieve the model from the server. Figure 14 shows the code that uses a model to control the robot movement. The code makes a robot turn right, showing a happy face on the screen and playing the "Meow" sound when it "sees" a face (the image classification classes are "face" and "no face"), and after 2 s, it shows a normal face on the screen.

The future version of the Scratch extension will include "text-to-speech" and "speech recognizer" functions.

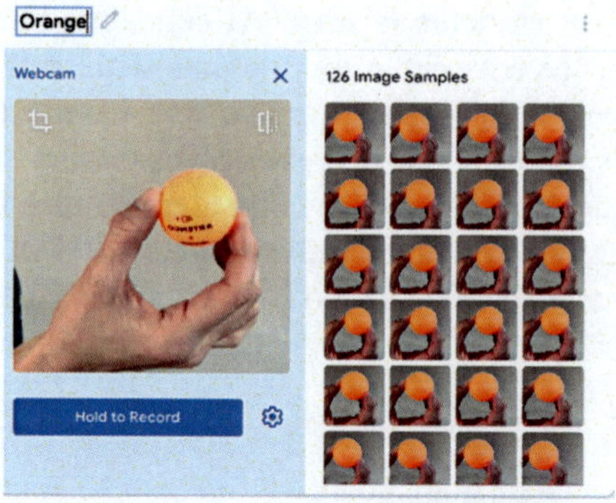

Fig. 11. Creating image samples

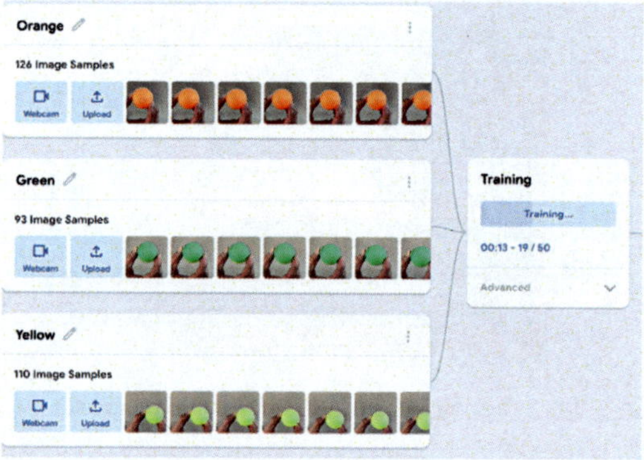

Fig. 12. Training an image classification model

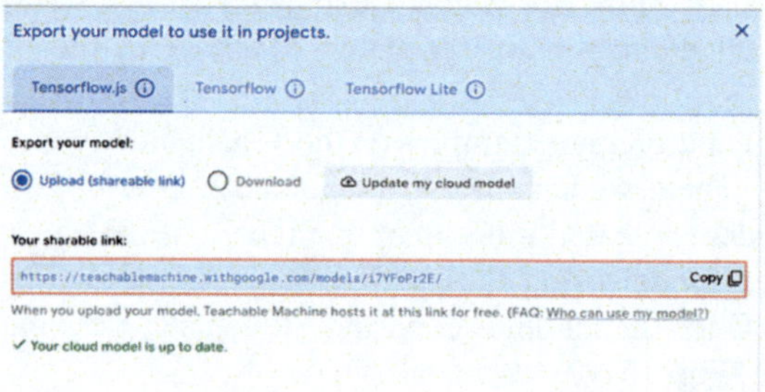

Fig. 13. Exporting a model

Fig. 14. Code using an image classification model

3.6 CogBot Lesson Ideas

CogBot lessons align with the 5 Big Ideas in AI introduced by the AI4K12 initiative. CogBot lessons introduce students to machine learning while exploring how CogBot "senses" or "sees" its environment and understands his surroundings (perception), what it senses/sees and how it uses the information presented (representation and reasoning), how it learns to be able to recognize the images (learning), and how the data gathered could influence its understanding of the surroundings and decisions (societal impacts) by training models. The lessons provide sample codes for students to tinker with, such as Peek-a-boo (Fig. 14) and Guess Who (Fig. 10) codes.

While training models, students are prompted to wonder how many images are enough for CogBot to accurately identify a face or object (explore data size). It introduces the concept of a "confidence threshold" to help them understand how it works in defining a face/object. When tinkering with the Guess Who code, students can train a model with multiple faces or objects to explore if CogBot can identify different faces or objects (exploring types of data). Then students are prompted to question if CogBot can identify an image that is not included in the model (i.e. their teacher's face) to explore how training data can influence the AI's performances (societal impacts).

Once students understand how to program CogBot with Teachable Machine, students can work on their own projects. One project could be to create a robot that can translate sign language. Students create sign language (or use an existing one) and train a model so that CogBot can recognize different signs. Students can explore how close or far a hand can be for ThinkBot Robot to be able to identify a sign correctly, and/or whether ThinkBot Robot can recognize a sign when wearing darker gloves or the background color is totally different to understand the limitation of AI (societal impacts). Another project could be to create a CogBot town, simulating their own town/city. Students can place different figures (or create figures), street signs, and buildings/objects where CogBot runs around autonomously. Students need to figure out how to make CogBot identify different signs, objects, and figures and autonomously run in the town successfully.

4 Next Step and Challenges

The paper introduced CogBots, an affordable, open-source, AI-enhanced educational robotics tool, for promoting AI literacy including the 5 Big Ideas in AI. They are affordable since the materials they use are less expensive and available in many countries. CogBots can be built with 3D printed parts by downloading CAD files or using craft materials and/or recycled cardboards and other materials available. CogBots are open-source and the materials for building and learning and teaching will be shared online. One of the next steps to develop a community website where CogBots materials and lessons are shared, and constant updates are made both by the developers as well as the community users.

Since the project aims to make AI-powered learning available for *everyone* in the world, there are several challenges for which solutions need to be developed. The current model of CogBots requires computers for programming and creating training models. With the next iteration, both the training and coding functions will be added to the application, which allows the students in the countries where access to computers is limited in schools to be able to learn with CogBots. This also allows students without a WiFi connection to be able to learn with CogBots. Moreover, it allows the phone to train models. This provides more opportunities for students to develop creative projects.

Currently making sure that there are enough Android phones available for a classroom is a challenge. It is a challenge even in the U.S. where iPhone is the most popular smartphone, especially among young people. Finding recycled Android phones enough to use in a classroom is quite challenging. There needs to be some system established where recycled Andriod phones, and iPhones, in the future, can be donated for use in schools. Although there are issues, CogBots can provide inclusive learning opportunities in various countries, especially equitable access to AI-powered educational robotics learning tools in developing countries where technological infrastructures are still underdeveloped.

Reference

1. United Nations Educational, Scientific and Cultural Organization (UNESCO): Artificial Intelligence for Sustainable Development - Synthesis Report: Mobile Learning Week (2019)
2. Higuera, C.D.: A report about Education, Training Teachers and Learning Artificial Intelligence: Overview of key issues (2019)
3. United Nations Educational, Scientific and Cultural Organization (UNESCO): Artificial Intelligence in Education: Challenges and Opportunities for Sustainable Development (Working Papers on Education Policy) (2019)
4. Touretzky, D., Gardner-McCune, C., Martin, F., Seehorn, D.: Envisioning AI for K-12: what should every child know about AI? In: Proceedings of the AAAI Conference on Artificial Intelligence, Honolulu, HI. AAAI Press (2019)
5. Bers, M.U., Seggighin, S., Sullivan, A.: Ready for robotics: bringing together the t and e of stem in early childhood teacher education. J. Technol. Teacher Educ. **21**, 355–377 (2013)

6. Eguchi, A.: Educational robotics for elementary school classroom. In: Proceedings of the Society for Information Technology and Education (SITE), pp. 2542–2549 (2007)
7. Eguchi, A.: Educational robotics as a learning tool for promoting rich environments for active learning (REALs). In: Keengwe, J. (ed.) Handbook of Research on Educational Technology Integration and Active Learning, Hershey, PA, pp. 19–47. IGI Global (2015). Information Science Reference

Social Robots

The Socially Assistive Robot Daisy Promoting Social Inclusion of Children with ASD

Sofia Pliasa[1,2], Anna-Maria Velentza[1,2], and Nikolaos Fachantidis[1,2](✉)

[1] Department of Educational and Social Policy, University of Macedonia, Thessaloniki, GR, Greece
{spliasa,annamariavel,nfachantidis}@uom.edu.gr

[2] Laboratory of Informatics and Robotic Applications in Education and Society, University of Macedonia, Thessaloniki, GR, Greece

Abstract. Children in the autism spectrum face deficits in communication's social skills, which makes their inclusion difficult to obtain. Socially Assistive Robots, according to recent researches, have proven to be capable tools of assisting children with ASD to socialize and motivate them to participate in group activities. The research of this study explores whether Daisy Robot, a socially assistive robot, can motivate children with ASD to verbally communicate with peers of typical development by developing verbal communication skills, to that extent that those skills are generalized after the interventions while interacting with different children of typical development. Six children with ASD were observed previously, during, and after an intervention with the Daisy robot. All six managed to develop and maintain their verbal communication skills towards their social inclusion.

Keywords: Socially assistive robots · Daisy robot · Social inclusion · ASD children

1 Introduction

An increasing number of children worldwide are being diagnosed with autism spectrum disorders (ASD) [1–4]. Based on recent studies, 1 out of 59 children are on the autism spectrum (CDC, 2017). Children with ASD usually experience different, mild, or severe expressions of the disorder. Some may have a normal or high IQ, while others may have mental disabilities. They also tend to respond impulsively to sensory stimuli, interact with objects in strange ways, or be extremely attached to them [5].

Based on "Diagnostic and Statistical Manual of Mental Disorders, 5th edition" (DSM-V), the diagnostic criteria of the autism spectrum are identified in the following:

1. deficits in communication and social and emotional reciprocity. More precisely:
– Attempts at social approach in unusual ways, difficulty, or even failure to participate in a discussion, reduced perception of emotions, general difficulty in responding to social interactions.

M. Malvezzi et al. (Eds.): EDUROBOTICS 2021, SCI 982, pp. 89–102, 2021.
https://doi.org/10.1007/978-3-030-77022-8_8

- Deficiencies in verbal and non-verbal communication, ranging from poor vocabulary, lack of eye contact, and insufficient understanding of body language and gestures, to even the absence of speech and expression.
- Difficulties in maintaining and understand social relationships, such as adapting their behavior to a variety of social conditions, difficulties in symbolic play, and lack of interest in peers.

2. Restricted or repetitive patterns of behavior, interests or activities in at least two of the following categories, either present or past:

- Persistence in similarity, rigid adherence to routines, ritual motifs, or even stereotypes of verbal and non-verbal behaviour.
- Extremely limited, specific interests that are abnormal in intensity or focus.
- Hyperactivity or atony in sensory perception or unusual interest in sensory stimuli of the environment (DSM-V).

ASD is divided according to the severity of the symptoms to:

- Level 3 – «Need for Special enhanced support» (serious difficulties in socialization and flexibility),
- Level 2 – «Need for enhanced support» (important difficulties) and
- Level 1 – «Need for support» (difficulties).

Socially assistive and educational robots can enhance special education students to practice and learn practical and cognitive skills, like collaboration, and even gain self-confidence [6]. The use of Intelligent Robot Agents with humanoid characteristics ('AnRI') has been already used in Special Education [7] and Kaspar robot is successful at performing autonomous interactive tasks with children with special needs safely and reliably [8]. Similarly, for students in typical education, socially assistive robots have proven beneficial in tutoring skills in a similar manner with human tutors on restricted tasks and manage to successfully increase students' affective and cognitive out-comes [9]. Additionally, Neumann supports the idea that children- social robot interaction can enhance their language and communication skills [10].

Inclusion is a system that accepts diversity as a rule, providing high quality education with a "personalized approach based on each student's needs" [11]. Successful inclusion is achieved in environments that promote social interaction and provide social interaction opportunities, which encourage children to engage in common activities and act in a socially acceptable way. [12, 13]. The successful inclusion of children with ASD presupposes that they should be able to:

- participate more in joint activities and interact socially,
- be acceptable and surrounded with social support,
- have friendship networks,
- achieve more advanced educational goals in comparison to children attending special schools [14–16].

The development of relationships between children with ASD and their peers is often challenging. An additional reason stems from the limited ASD students' skills to start and complete a group game due to the lack of adaptation to the games' rules and new conditions [17, 18]. Children with ASD need teaching and guidance before they can interact with others [19, 20].

Independent studies in different contexts show that interventions with social assistive robots can help children with ASD to engage in activities that require collaboration and communication skills. Social assistive robots are commonly used from children to communicate, express and perceive emotions, maintain social relationships, interpret physical stimuli and develop social skills [21, 22].

All those are used as tools for teaching skills to children with autism, to play with them and to provoke from them desirable behaviours [23–25]. They create interesting, attractive, meaningful conditions that motivate children to interact with them. Moreover, various robots have been created to fulfill that purpose, which vary in appearance, behaviour, and performed activities [26], such as Kaspar, Milo and Keepon [27].

2 Current Study

The purpose of the current study is to explore whether an intervention with the social assistive robot Daisy can help children with ASD to develop social skills of verbal communication while interacting with peers of typical development (TD) and whether those skills can be generalized in follow up sessions with different peers of TD.

2.1 Communication Skills

Although the terminology varies between different researches, the main communication skills studied are specific. According to the literature, four main social-communication deficits are distinguished:

- **verbal communication intentions** which are defined as comprehensible vocabulary addressed to adults or peers for the purpose of attracting attention, commenting, answering as well as asking a question,
- **verbal responses** which are defined as the verbal recognition of the onset of verbal communication by an adult or peer and which relates to comprehensible vocabulary,
- **focus of common attention** which refers to the child's ability to show, point, share a smile or look with an object and with an adult or peer,
- **non-verbal communication attempts** which are defined as gestures, and the clear use of body language that do not involve the focus of common attention, and operate on your non-verbal comments or requests.

These are also considered to be the most obvious indicators of the differences, in social language, between children with ASD and the formal development of their classmates [28].

3 Experimental Design

To test the hypothesis of this research the scores of children with ASD on the aforementioned skills, the scores of the children were recorded and where compared prior (BS

- Baseline Session), at the end of a three steps intervention (FS – Final Session), and almost after a month in a follow up session (FU).

3.1 Participants

Eighteen children, five to nine years old participated in the particular intervention, six of them with ASD (children K1, K2, K3 that were kindergarten students and children E1, E2, E3 that were elementary students), six of TD for the interventions and six of TD for the follow up session. All children were attending Greek public schools. For the session prior to the intervention a teacher also participated to introduce the games to the six groups of children and to guide them throughout its course. The parents of all children signed a consent, allowing their children to participate in the interventions.

Children with ASD. In the interventions participated, nine kindergarten students and nine elementary students. In the table below the groups of children that teamed up for the interventions are presented (Table 1). According to the inclusion criteria all the children with ASD could verbally communicate and according to their diagnoses (from Greek Children Diagnostic and Treatment centres) were in the autism spectrum at the level 1 of severity. They were highly functional but with significant difficulties and deficits in social skills. Data on their IQ were not available.

Table 1. Groups of participants

Group	Group 1		Group 2		Group 3		Group 4		Group 5		Group 6	
	Student	Age	Student	Age	Student	Age	Student	Age	Student	Age	Student	Age
Student with ASD	E1	7	E2	8	E3	9	K1	6	K2	6	K3	6
Student of TD	T1	7	T2	8	T3	9	T4	6	T5	6	T6	6
Student of TD follow up	FU1	7	FU2	8	FU3	9	FU4	6	FU5	6	FU5	6
Teacher	Teacher 1		Teacher 1		Teacher 1		Teacher 2		Teacher 3		Teacher 3	

3.2 Settings

The research took place in Kindergarten and Elementary public schools of Thessaloniki, Greece at which the children were attended. Specifically, the activities took place in the classrooms of specialized learning ("inclusion classrooms") that were quiet enough and without many sensory stimuli.

For the intervention we used a desk and three chairs that were placed in front of it for the three participants to be seated. For the sessions with the assistive robot, Daisy Robot was placed on the table, with its face looking at the participants. On the table was also a tablet and board games for the implementation of the games-activities. The observer sat at the corner of the room, away from the participants so that could operate the robot and keep notes without distracting the participants.

3.3 Materials

Daisy Robot was utilized in order to mediate interactions among children with ASD, typical development children and their teacher. Daisy robot, which is a SAR, was utilized for the implementation of the interventions. Daisy is a semi-autonomous robot in the shape of a flower that resembles a stuffed soft toy, with light blue and purple colour, as shown in Fig. 1. It can show emotions and facial expressions, simulate eye contact, and perform movements with the petals located around the body.

Fig. 1. Daisy robot

4 Data Collection - Measurement

In order to create the observation scale, a tool based on TEACCH's Social Assessment Method [29] was modified and utilized [30, 31]. After editing and modifying the tool, the following skill scales (markers) with their sub-scales were set for testing:

Information (IN). The ability of the child to give and ask for information about himself, objects and other people, the ability of answering questions using one word or a clarifying way.

Discussion skills (DS). The child's ability to talk, narrate.

The scales and subscales that were observed are presented below (Table 2).

Table 2. Skill scales-subscales

Information
1a The child's ability to provide information about him/herself
1b The child's ability to provide information about others
1b The child's ability to provide information about others
Discussion skills
2a The child's ability to address his/hers attention to the person that speaks
2b The child's ability to retain acceptable distance from his/hers teammate
2c The child's ability to describe past experiences

The TEACCH observation protocol [29] was used to grade the skills. According to the protocol, children's behaviors are assessed by means of appropriate responses and whether the child can respond independently or when prompted (Table 3).

Table 3. Olley's observation protocol

Social skill	Level of autonomy				
Child's response	Inadequate persistent response	Physical prompt requires	Special prompt required	General verbal prompt required	Appropriate response without prompting
Interpretation	Refuses to respond (participate) even with a physical prompt	It responds satisfactorily to a physical stimulus	Responds satisfactorily to a special prompt (e.g. "George, go to a seat")	General verbal prompt required ("all children sit in yours seats")	No prompt is required (follows instructions independently)

In order to ensure the credibility of the study, the researcher and the teacher were both taking notes on the observation scale. The notes concerned the way children were responding to the conversation routine, their participation in the game activities, and what was the frequency of the observed behaviours. After the end of each session, both the observer and the teacher compared their notes and their skills were graded with the agreement of both of them.

4.1 The Intervention

Baseline Session. Prior to the intervention, the two children in each group (child with ASD and TD child that participated in the intervention), under the guidance of the teacher, were asked to answer questions about themselves and their family, to state favourite objects and habits, and to engage in dialogue with their teammates. In addition, the children were asked to describe experienced events of their everyday lives. Throughout the meeting it was observed were the attention of children with ASD was oriented and whether they kept an acceptable distance from their classmates.

The children's scores were kept to be compared at the end with the scores of the final and follow up evaluation.

Steps of the intervention. A three-step intervention was designed and implemented. Before the actual intervention two sessions took place as a familiarization phase to the robot. During the first session the interaction between the Daisy Robot and the child with ASD was initiated and established with conversation and game activities and during the second session the teacher entered the team. The presence of the teacher ensured the smooth conduct of the games, as the teacher knew the child's habits and what was likely to irritate or help him/her to work better, so as possible anxiety triggers to be avoided. After this familiarization phase the actual intervention took place.

Step one. The main objective of this step was the acceptance of the child of typical development to be achieved by the child with ASD and its incorporation into the team (teacher – child with ASD- Robot) to be succeeded. During this step both children and the teacher followed the protocol of the baseline session. The child with ASD was encouraged by the robot to address questions to his/hers teammates, and to try to establish dialogue patterns. The robot initiated and guided the interactions, provided with the instructions and the necessary prompts for the successful implementation of the procedure. The teacher's role in this step was to help the child with ASD to feel comfortable and safe in the new context as the child of TD is more spontaneous in his/her reactions and therefore unpredictable, which can cause anxiety to the child with ASD, who may then be unable to cooperate properly.

Step two. During step 2, the child's in the spectrum reference point - the teacher is absent. The aim of this step is the cooperation and communication skills that have been achieved previously to be maintained, but with the minimal supportive presence. The robot does not without the full supportive frame that the teacher offered, still is a "significant other" but to a lower degree. The two children with the guidance of the robot followed the procedure of the previous steps.

Step three. In this step, the robot waived the lead role. It retired, declared tired and stayed a simple viewer of the interaction between the two children. The significance of this step was to realize whether any skills required in previous steps, were maintained besides the absence of the robot. step, was is important to see whether the skills acquired in the previous were maintained in a context of equal coexistence, without guidance from a significant other, but through communication and cooperation with the co-worker (teammate). The teacher initiated the questions and the children were observed.

5 Results

To answer the research's hypothesis, the Wilcoxon signed rank test (at 95% confidence level) was applied to compare the performances of the children with ASD in all two components that were tested: information (I) and discussion skills (DS), during the BS, FS, FU sessions.

At Table 4 we can see the scores of each six children with ASD at BS, FS and FU at the skill of information. As it is observed all children had better scores during the FS which was the last step (step 3) of the intervention. During the FU the scores of the children were less high that the FS but were still retained higher that scores of the baseline session (BS). At the diagram (Fig. 2), all the children's with ASD scores, as a total, are observed at the three evaluation points (BS, FS, FU) and as it is shown the six children performed better at the FS. Table 5 shows that there is a statistically significant difference in the skill of information between the Baseline (BS) and the Final session (FS) and between the Follow Up session (FU) and the Final session (FS).

Similar are the results of the Skill of Discussion at it observed at the tables and the figure below (Tables 6–7, Fig. 3).

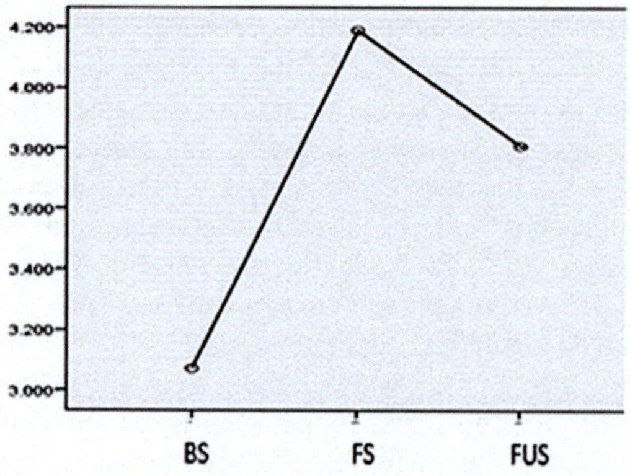

Fig. 2. Diagram of skill of IN

Table 4. Scores of skills of IN

	Information		
	BS	FS	FU
K1	2,571	4,142	3,428
K2	3	4	3,571
K3	2,142	3,285	2,714
E1	4,571	4,857	4,714
E2	2,285	4,142	3,714
E3	3,857	4,714	4,142

Table 5. Test Statistics c skill of IN

	I_FS – I_BS	I_FU – I_FS
Z	– 2,201[a]	– 2,201[b]
Asymp. Sig. (2-tailed)	,028	,028

a. Based on negative ranks
b. Based on positive ranks
c. Wilcoxon Signed Ranks Test

Table 6. Scores of SD

	Skills of discussion		
	BS	FS	FU
K1	1,75	4	3,25
K2	2,375	3,25	2,75
K3	1,375	2,5	1,75
E1	3,75	4,75	4,25
E2	3,375	4,375	4
E3	3,375	4,75	3,625

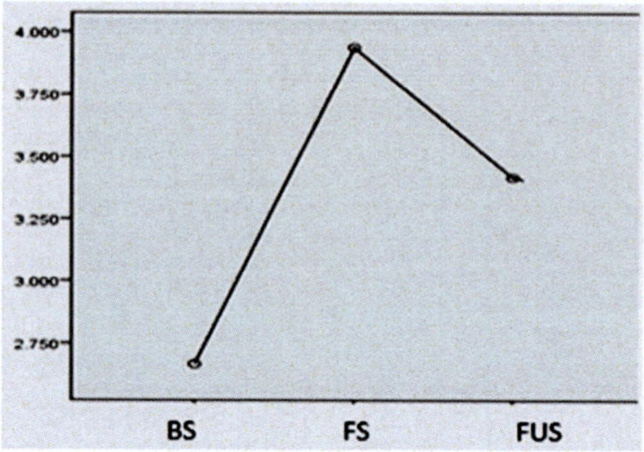

Fig. 3. Diagram of skill of IN

6 Discussion and Conclusion

The goal of this study was to determine whether a session with a robot as a facilitator can effectively assist with children with ASD to develop social skills of communication and whether those skills can be maintained so as to be transferred while interacting under different circumstances so as to promote the inclusion of children with ASD.

Children seemed to more willingly participate in the dialogue patterns, answer questions will less hesitation and seem more interested in the interaction with their peers when guided by the robot, conclusion that agrees with previous researches' outcomes, that suggest that triadic interactions (child with ASD-robot-another human) can be successfully fostered when initiated by a robot [32, 33]. Children with ASD showed increased interest in verbally communicating, providing information about themselves and even initiating a conversation, also as previous researches verify which suggested that robots can lead to a self-initiated social interactions with a human partner [34, 35] and also to show more engagement [36].

Furthermore during the initial interaction to the robot the enthusiasm and security that children with ASD felt by its presence, motivated them to approach and interact with Daisy robot which verifies the surveys that present SAR able to engage children in the autism spectrum in collaborative activities [37, 38].

In conclusion, Daisy robot seems to be able to provide consistent, repeatable behaviors that promote a stable environment, in which children with ASD feel secure and confident to fulfil verbal interactions. The skills of Information and Discussion were maintained while the children with ASD interacted with different peers thus their inclusion was promoted.

The limited sample of six children with ASD does not allow great generalization of the results. Significant data will result from a research in a large heterogeneous sample, the analyses of which will prove the added value of Daisy Robot in the development of social skills in children with ASD.

Acknowledgments. This research is co-financed by Greece and the European Union (European Social Fund- ESF) through the Operational Programme «Human Resources Development, Education and Lifelong Learning 2014–2020» in the context of the project "Mapping the characteristics of socially assistive robots with the aim of enhanced cognitive functions and intimacy in humans. "MIS 5047258".

Appendix

Example of Step 2 Dialogues between the child in the autism spectrum (E2), child in typical development (T5) and Daisy robot.

E2 and T5 enter the room and sit at the chairs in front of Daisy. E2 shows no hesitation at all and is laughing.

Daisy:	Hello E2 (name of the child).
E2:	Hello.
Daisy:	Is this your friend?
E2:	Friend T5 (name of the child of TD)
Daisy:	Hello T5, I am Daisy. How are you?
T5:	Hi Daisy, nice to meet you, I am doing good.
Daisy:	E2 can you ask your friend what is her favorite color?
E2 (continues to look at Daisy and repeats):	What is her favorite color?
Daisy:	E2 look at your friend
	E2 looks at her
Daisy:	ok, now ask your friend "What is your favorite color?"
E2:	What is your favorite color?
T5:	I think orange is my favorite color
Daisy:	Nice color, I love orange. E2 can you please remind me what is your favorite color?
E2:	Red
Daisy:	I love red too. What is your dad's name?
E2:	Dad name Tom
Daisy:	ask your friend what her favorite food is.
	E2 looks at T5 and repeats: what is her favorite food?
Daisy:	E2 ask her "What is your favorite food?"
	E2 (repeating Daisy's questions): What is your favorite food?
T2:	I love eating fish.
Daisy:	E2 do you like eating fish?
E2:	Fish. (E2 moves his head left to right like saying no.)
Daisy continues with some more questions and then says:	Let's play a game, do you want to play a game?

T5:	I'd love to E2 doesn't respond but grabs the tablet and looks at it.
Daisy repeats the instructions:	T5 and E2, here is a flower, you need to drag one petal at a time, first E2 and then T5. E2 starts and then waits for his friend; T5 drags a petal and then E2 starts dragging two petals. Daisy corrects him
Daisy:	E2 you drag only one petal, then your friend. E2 obliges. E2 is excited and they repeat the game two more times.

References

1. Blaxill, M.F.: What's going on? The question of time trends in autism. Public Health Rep. **119**, 536–551 (2004)
2. Olds, J., et al.: Implications: human cognition and communication and the emergence of the cognitive society. In: Roco, M., Bainbridge, W., Tonn, B., Whitesides, G. (eds.) Convergence of Knowledge, Technology and Society (2013)
3. Scassellati, B., Admoni, H.: Mataric, M: Robots for use in autism research. Ann. Rev. Biomed. Eng. **14**, 275–294 (2012)
4. Wong, et al.: Evidence-based practices for children, youth, and young adults with autism spectrum disorder. Frank Porter Graham Child Development Institute University of North Carolina at Chapel Hill (2014)
5. Feinstein, A.: A History of Autism: Conversations with the Pioneers. Wiley-Blackwell, London (2010)
6. Karna-Lin, E., et al.: Can robots teach? Preliminary results on educational robotics in special education. In: Sixth IEEE International Conference on Advanced Learning Technologies (ICALT 2006), pp. 319–321 (2006)
7. Zahariev, R., Valchkova, N., Wagatsuma, H.: Service robots for special education of children with disabilities: robotized systems for social applications. In: Proceedings of the 21st International Conference on Computer Systems and Technologies, New York, NY, USA, pp. 300–306 (2020)
8. Wood, L.J., et al.: Developing kaspar: a humanoid robot for children with autism. Int. J. Soc. Robot., 1–18 (2019)
9. Belpaeme, T., et al.: Social robots for education: a review. Sci. Robot. **3**(21), eaat5954 (2018)
10. Neumann, M.M.: Social robots and young children's early language and literacy learning. Early Child. Educ. J. **48**(2), 157–170 (2020)
11. Irvine, A.N., Lynch, S.L.: Inclusive education and best practice for children with autism spectrum disorder: an integrated approach. Int. J. Inclusive Educ. **13**(8), 845–859 (2009)
12. Zakai-Mashiach, M., Dromi, E., Al-Yagon, M.: Social inclusion of preschool children with ASD: the role of typical peers. J. Spec. Educ. **10**, 0022466920926132 (2020)
13. Goodman, G., Williams, C.M.: Interventions for increasing the academic engagement of students with autism spectrum disorders in inclusive classrooms. Teach. Except. Child. **39**(6), 53–61 (2007)
14. Roberts, J., Webster, A.: Including students with autism in schools: a whole school approach to improve outcomes for students with autism. Int. J. Inclusive Educ., 1–18 (2020)

15. Fryxell, D., Kennedy, C.H.: Placement along the continuum of services and its impact on students' social relationships. J. Assoc. Persons Severe Handicaps **20**, 259–269 (1995)
16. Hunt, P., et al.: Evaluating the effects of placement of students with severe disabilities in general education versus special classes. J. Assoc. Persons Severe Handicaps **19**, 200–214 (1994)
17. Kasari, C., Patterson, S.: Interventions addressing social impairment in autism. Curr. Psychiatry Rep. **14**(6), 713–725 (2012)
18. Kasari, C., Rotheram-Fuller, E.: Peer relationships of children with autism: challenges and interventions (2007)
19. Berry, A., et al.: Use of assistance and therapy dogs for children with autism spectrum disorders: a critical review of the current evidence. J. Altern. Complement. Med. **19**(2), 73–80 (2013)
20. Watt, N., et al.: Repetitive and stereotyped behaviors in children with autism spectrum disorders in the second year of life. J. Autism Dev. Disord. **38**(8), 1518–1533 (2008)
21. Cabibihan, J.J., et al.: Why robots? A survey on the roles and benefits of social robots in the therapy of children with autism. Int. J. Soc. Robot. **5**(4), 593–618 (2013)
22. Ismail, L.I., et al.: Leveraging robotics research for children with autism: a review. Int. J. Soc. Robot. **11**(3), 389–410 (2019)
23. Abbasi, J.: In-home robots improve social skills in children with autism. Jama **320**(14), 1425 (2018)
24. Yun, S.S., et al.: Social skills training for children with autism spectrum disorder using a robotic behavioral intervention system. Autism Res. **10**(7), 1306–1323 (2017)
25. Kim, E.S., et al.: Social robots as embedded reinforcers of social behavior in children with autism. J. Autism Dev. Disord. **43**(5), 1038–1049 (2013)
26. Huijnen, C.A., et al.: Mapping robots to therapy and educational objectives for children with autism spectrum disorder. J. Autism Dev. Disord. **46**(6), 2100–2114 (2016)
27. Yousif, J.H., Kazem, H.A., Chaichan, M.T.: Evaluation implementation of humanoid robot for autistic children: a review. Int. J. Comput. Appl. Sci. **6**(1), 412–420 (2019)
28. Murdock, L.C., Cost, H.C., Tieso, C.: Measurement of social communication skills of children with autism spectrum disorders during interactions with typical peers. Focus Autism Dev. Disabil. **22**(3), 160–172 (2007)
29. Olley, J.G: The TEACCH curriculum for teaching social behavior to children with autism. In: Social Behavior in Autism, pp. 351–373. Springer, Boston (1986)
30. Watson, L.R., Marcus, L.M.: Diagnosis and assessment of preschool children. In: Diagnosis and Assessment in Autism, pp. 271–301. Springer, Boston (1988)
31. Schopler, E., Mesibov, G., Baker, A.: Evaluation of treatment for autistic children and their parents. J. Am. Acad. Child Psychiatry **21**(3), 262–267 (1982)
32. Pliasa, S., Fachantidis, N.: Using daisy robot as a motive for children with ASD to participate in triadic activities. Themes eLearning **12**(12), 35–50 (2019)
33. Stanton, C.M.: Robotic animals might aid in the social development of children with autism. In: 2008 3rd ACM/IEEE International Conference on Human-Robot Interaction (HRI), pp. 271–278. IEEE (2018)
34. Pliasa, S., Fachantidis, N.: Can a robot be an efficient mediator in promoting dyadic activities among children with autism spectrum disorders and children of typical development? In: Proceedings of the 9th Balkan Conference on Informatics, pp. 1–6 (2019)
35. Robins, B., Dautenhahn, K.: The role of the experimenter in HRI research—a case study evaluation of children with autism interacting with a robotic toy. In: Proceedings of the 15th IEEE International Symposium on Robot and Human Interactive Communication, pp 646–651. IEEE Press, Piscataway (2006)

36. Robins, B.: Robotic assistants in therapy and education of children with autism: can a small humanoid robot help encourage social interaction skills? Univ. Access Inf. Soc. **4**, 105–120 (2005)
37. Pliasa, S., Fachantidis, N.: Mobile technologies serious games for the development of social skills in children with autism spectrum disorders, in enhanced with socially assistive robots interventions. In: Interactive Mobile Communication, Technologies and Learning, pp. 618–628. Springer, Cham (2019)
38. Robins, B., Dautenhahn, K., Dickerson, P.: From isolation to communication: a case study evaluation of robot assisted play for children with autism with a minimally expressive humanoid robot. In: Proceedings of the Second International Conferences on Advances in Computer-Human Interactions, ACHI 2009 (2009)

Edù, a Robotic Companion in Pediatric Protective Isolation Units

Filippo Talami(✉), Maximiliano Romero, and Giovanni Borga

Università IUAV di Venezia, 30135 Venice, VE, Italy

Abstract. Hospital isolation is often a traumatic and stressful experience, especially for children, since it implies not only the acceptance and management of the illness but also the detachment from the family environment. This paper presents a methodology to develop a social robot to accompany children in a critical period of medical care: the protective isolation units. A user-centered and multidisciplinary design approach has been adopted to include not only the technological aspects of robot development but also the psychological impact of its adoption. Desk research highlighted the technical limitations of pediatric isolation, while medical experts offered insights on the medical procedures. In addition, applied research with flexible materials was conducted during the development of the final prototype. As a result, a pediatric companion robot called Edù was developed to help hospitalized children coping with their daily routine and teach them how to behave during medical procedures and checkups.

Keywords: Design education · Robotic companion · Hospital isolation

1 Introduction

Recently, social robots have become more and more present in our world and, with the development of new technologies, they are becoming able to help people in difficult situations. Although most of the social robots commercially available seem not to be designed for a specific field of implementation, they are increasingly being developed, tested, and implemented in the healthcare sector [1–3], such as the care of elderly people, nurse assistance, and hospital check-ins. However, research on the application of social robots designed to help children in the healthcare context appears to be at an emergent stage [1–4]. Additionally, it seems that there is a lack of social robots designed specifically for children being in pediatric isolation due to chronic diseases or infections [5]. For a child, being in hospital isolation, which can last up to one month, is a traumatic and stressful experience, which implies not only the acceptance and management of their illness but also the separation from loved ones and the family environment.

A robotic companion could be very helpful in these situations due to its ability to gather updated medical data, acquired during games sessions with the child, by being hugged or through daily conversations. A robot assistant could also teach and explain how to behave during medical checks and during operations or help children to cope with the situation before or after each medical procedure.

M. Malvezzi et al. (Eds.): EDUROBOTICS 2021, SCI 982, pp. 103–107, 2021.
https://doi.org/10.1007/978-3-030-77022-8_9

2 Material and Methods

2.1 Information Acquisition and Context Definition

Before the design process, data acquisition was carried out mostly through scientific papers [6, 7], surveys filled by stakeholders, and discussions with experts in the field. Giulia la Placa from the Pediatric Hematology department of Torino hospital has been interviewed to understand user's needs and fears. Dr. Cesaro from the Pediatric Hematology and Oncology department of Verona hospital gave insights on the patient's routines and some of the hospital internal procedures.

After gathering the data, it has been identified that children isolating due to very infectious diseases or immunodeficiency needed help to cope with the situation while ensuring mental health during procedures before surgery, or post-surgery treatment. In fact, in this situation, children need special attention regarding education and social activities, but no human can stay in the same room for more than a few hours without endangering themselves.

2.2 Context Description and Internal Procedure Rules

There are several types of hospital isolation procedures, applied in different ways and at different levels, to prevent the spread of infectious contaminants and maintain sterile spaces with restricted access. A room may have either negative or positive pressure in order to isolate patients with infectious conditions or to keep vulnerable patients safe from infections and disease. Objects such as toys made of synthetic fur or organic material for hygiene purposes cannot be brought inside the room, and if accepted must be for exclusive use of the resident. As for the visits, only one relative can be admitted each day and must be using personal safety protection devices. Visitors are prohibited from entering in case of cold or flu symptoms, as well as people who had contacts with children affected by infectious diseases, including sisters or brothers attending school.

2.3 Brief and Concept

The insights gained during the desk and field research highlighted that a social robot designed to work in pediatric isolation should provide three main features:

- Helping the children coping with the situation.
- Collaborating with the medical staff gathering medical and psychological data.
- Acting as a means of communication between children, family, and medical staff.

The embodiment of the robot should communicate warmth and lightness and should provide the possibility to be hugged by a child. A social robot should also be expressive using its eyes and mouth to simulate mood and personality. Voice communication should be used to help the children talking about fears and issues and relax before surgical procedures. Table 1 breakdowns the design solutions to tackle the highlighted limitations through the design of a social robot for pediatric isolation.

Table 1. Limitations of pediatric isolation and design solutions through a social robot.

Limitation	Solution
1. A child must spend all day in the bedroom	1. The social robot helps to break daily routine
2. Sometimes it is hard to get a child to cooperate	2. It mediates between kids and doctors
3. It is necessary to comfort the patient after medical procedures	3. It explains medical procedures and remember past events
4. Every item must be sanitized	4. It withstands cleaning and sanitizing procedures
5. For faster healing, the child should feel useful to someone	5. It requires some help from the patient for moving

3 Results

According to the well-known User-Centered Design (UCD) approach, several prototypes have been made to better assess user interaction and improve the final goals. The first prototype was made in polystyrene and clay to estimate proportions and size. Subsequently, a 3D printed model and a silicone sock helped to evaluate embodiment and appearance. The working prototype aided to test user interaction adopting an embedded Arduino Mega board that controlled two LCDs for the eyes of the robot, two speakers for its voice while three LEDs were adopted to simulate facial expressions. It also included 4 touch sensors, each one triggering a different reaction.

Edu's working prototype required three months of iterative processes focusing on user tests and improvements. Figure 1 depicts outcomes of the design process from the initial model until the final prototype. Edu's latest version is about 30 cm high and only 1 kg heavy, making it easier to lift and place on top of a hospital bed. Edu's body is made of polypropylene with a shell of polyurethane foam and a Liquid Silicone Rubber (LSR) sleeve. In fact, LSR is a water and dustproof material, therefore ensuring the safety of the electromechanical components and it is completely sterilizable with current methods applied in hospitals.

The final product will have a machine learning programmed AI able to process the signals of an HD micro camera and 4 microphones, recognize emotional states and store information into a local and/or remote database that can be accessed by medical staff. Edù mainly interacts through conversations and games to help children coping with the situation as well as during their daily routine in the hospital between medical checks and preparation before medical procedures. Since daily routine in the isolation department can be very stressful for the child, the AI can recognize the child's emotions and act accordingly to relieve stress and fatigue. With an audio/video interface, Edù can also become an effective communication tool with the family allowing parents to play with their child through the robot when they are apart.

A fundamental part of helping the child accepting the situation is the importance of understanding the care process. For this reason, Edù comes with its wheelchair which acts as a means of moving and as a charging base for the robot to create connection and empathy with the child. This act of taking care of someone else helps the patient to get distracted and speed recovery. In fact, the wheelchair is programmed to stop at times or proceed in the wrong direction, showing that Edù has not yet mastered its ability in moving correctly using its wheelchair and so in need of help, inviting the child to actively participate in the learning process.

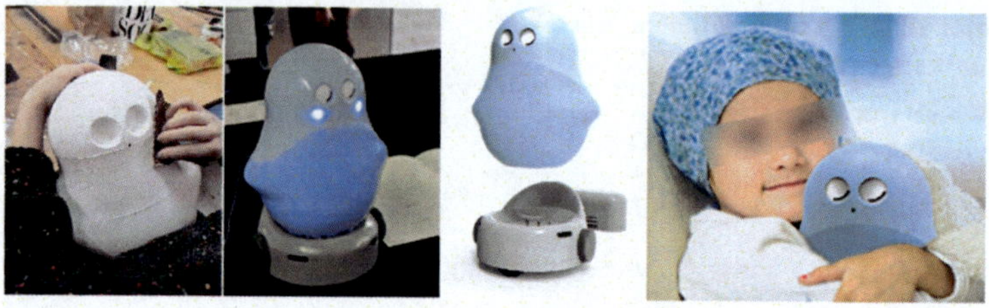

Fig. 1. The design process of Edù.

4 Conclusion

This paper showed that the development of a pediatric companion robot aimed at helping children, their families, and medical staff during hospital isolation is not a purely technological matter, but it also requires an iterative process based on tests with real users according to the UCD approach. This research is based on a problem setting stage carried out in strict collaboration with medical experts and on a robust test phase conducted using a working prototype. This work shows the benefits of adopting a multidisciplinary approach between designers and healthcare workers, the former skilled in setting the problem and the design process, the latter in handling and treating user issues. Further improvements will include multi-user management capabilities (useful in Covid-19 departments) and the implementation of additional interactions.

Acknowledgements. Edù is a social robot designed by Filippo Talami, Martina Bresciani, Gaston Brobjerg, Chiara Carucci, Mariacristina Nardone, students of the Product and Visual Communication Design Master at IUAV University. The authors would like to thank Dr.Cesaro, Marco Nalin, Giulia La Placa, and Denny Roncolato.

References

1. Breazeal, C.: Social robots for health applications. In: 33rd Annual International Conference of the IEEE Engineering in Medicine and Biology Society, Boston, MA (2011)
2. Dahl, T.S., Boulos, M.N.K.: Robots in health and social care: a complementary technology to home care and telehealthcare? Robotics **3**(1), 1–21 (2014)
3. Oborn, E., Barrett, M., Darzi, A.: Robots and service innovation in health care. J. Health Serv. Res. Policy **16**, 46–50 (2011)

4. Sharkey, A., Sharkey, N.: Children: the elderly, and interactive robots In: IEEE Robotics & Automation Magazine, vol. 18. IEEE, New York (2011)
5. Dawe, J., Sutherland, C., Barco, A., Broadbent, E.: Can social robots help children in healthcare contexts? A scoping review. BMJ Paediatr. Open **3**(1), 1–16 (2019)
6. Saldien, J., Goris, K., Vanderborght, B., Vanderfaeillie, J., Lefeber, D.: Expressing Emotions with the Social Robot Probo. Int. J. Soc. Robot. **2**, 377–389 (2010)
7. Bartneck, C., Forlizzi, J.A: Design-centred framework for social human-robot interaction. In: R&D Management (2004)

Development of a Robotic Agent for Increasing Elderlies Socialization

Giada Francescato(✉), Maximiliano Romero, and Giovanni Borga

Università IUAV di Venezia, Venice, Italy

Abstract. Social isolation is an increasingly felt problem in the elderly population and can significantly affect life quality. This paper describes the development process of a social robot that helps socialization among the elderly. The applied process consists of 5 main steps: desktop and literature research, user field research (observation and expert interviews), opportunity definition, and solution conceptualization, followed by a prototype development phase and a user tests phase. As a result, a companion robot called TIP (Therapy for Isolated People) was developed to help seniors participate in socializing activities by building a stable social network.

Keywords: Social isolation · Robotic education · HRI

1 Introduction

In recent years we have witnessed a wide adoption of social robots within home settings to provide support socializing with people. Assistive social robots, a particular type of robotics designed for social interaction with humans, could play an essential role in the health and psychological well-being of the elderly [1]. According to the 2019 EPTA European Parliamentary Technology Assessment report, it is estimated that by 2030, older people (60 years and over) will make up more than 25% of the population in Europe [2], as a consequence of the increase in life expectancy and the improvement in living conditions [3]. Among these, around 5 million older people live alone due to exclusion from social interactions, transforming loneliness into an increasingly felt and growing problem [4], which results in physical and mental health problems, including higher mortality rates, depression, and cognitive decline [5]. These data underline the necessity to develop new approaches to support them and re-educate them to social life, making aging healthy and active. This project aims to help older people establish new human relationships by encouraging their participation in events or meetings and creating a broad and reliable social network.

2 Materials and Methods

The project has been developed through five phases: desktop research with analysis of the context and state of the art, field research with interviews, concept, and prototyping. Finally, a brief evaluation with final users has been done.

M. Malvezzi et al. (Eds.): EDUROBOTICS 2021, SCI 982, pp. 108–111, 2021.
https://doi.org/10.1007/978-3-030-77022-8_10

2.1 Desktop and Field Research

Through an in-depth analysis of state of the art, it was found that social robots have positive effects on the elderly. Two types of robots can be distinguished: robots that assist in performing physical activities and robots that focus on the psychological well-being of the elderly [6]. Two socials educators from the San Lorenzo Service Center of Venice, one social educator of the Mini Accommodation Service of Venetian Public Institutions of Assistance (I.P.A.V.), and a group of six self-sufficient older people have been interviewed. The collaboration with these figures was fundamental to set up the right problem setting. The main findings in this phase concern the growing need for ways to deal with loneliness, the need for companionship, and the increasingly widespread habit of going out carrying personal technological devices. Other useful information were obtained from some tactile perception tests aimed at identifying the most suitable consistencies, textures, and colors for the coatings' production.

2.2 Concept

The project aims to provide an active interaction between the user and the robot. For this reason, it was deemed useful to exploit senses such as touch and hearing, equipping the robot with a voice of its own and the possibility to be touched to allow its activation favoring physical contact. The use of sounds and lights can help to draw the user's attention, motivating him to go out by listening to sounds from the outside or sounds that recall special activities. In addition to the main unit, a portable connected device can act as an emergency tool in case of need and as a sign of recognition once you reach the place where a scheduled activity occurs.

2.3 Prototyping

The prototyping phase involved the creation of various study models up to the final prototype. The first model was made using clay and was used to understand the size of the product better. Clay is a malleable material that made it possible to simulate the petals and experience movement. Two more 3D printed models of the robot body were made with Polylactic acid (PLA). The first model was made in white PLA, while the next model was made in transparent PLA after noticing that the light had a better yield. For the petals' development, tests were made to understand what material to make them and how the structure should be designed. Tests were made by building a cardboard petal, and then it was 3D printed with Thermoplastic Polyurethane (TPU). The final prototype, created to test the interaction and simulate reactions, uses a set of sensors, lights, motors, and a speaker controlled by an Arduino Mega board embedded in a 3D printed body covered in Lycra fabric.

2.4 Evaluation

The evaluation phase was carried-out by filming the user while interacting with the prototype to gather opinions on the acceptability and usability of the product. The feedbacks collected in this phase proved the effectiveness of light and movement as a reminder to the user, helping to locate the correct point to touch to activate the robot, as well as the use of sounds related to social activities in encouraging the user to go out. Furthermore, the user understood that having a partner to help him organize his day and participate in social activities helps him to stay active in and out of the house and to create new social networks.

3 Results

The whole project lasted about three months and resulted in TIP (Therapy for Isolated People), the social robot designed to help sociality among the elderly. TIP consists of two elements, the first one dedicated to the home and the second one portable device to use outside. The domestic device was designed to be placed on a surface; it requires constant energy and has a cylindrical body with a diameter of 115 mm and a height of 225 mm. The device (1) helps the elderly to stay in touch with what is happening in the neighborhood (events, news, sports) by stimulating conversation, (2) suggest participating in small daily activities by encouraging him to go out meeting people with the same interests, (3) helps to maintain contact with the new friends. The robot uses four modes of interaction: sound and voice, light, and the movement of the upper petals. Following a routine that develops gradually according to the user's habits, the device produces a sound of nature and light that attracts his attention. When the person touches the surface, the robot moves the petals and starts the conversation. Every day the device will propose different news and activities to be carried out outside the home. The robot perceives the touch through capacitive sensors distributed on the circular surface. Nylon cables connected to a central motor allow the movement of the petals. With a partial twisting of the engine, the wires are stretched, and the petals open and close. The outer part of the petal is made of TPU, a material with great flexibility and capable of deforming without losing its original shape, covered with an elastic fabric, the same that is used to cover the remaining surface of the robot. The portable device (1) helps the elderly in case of need by providing an emergency call button, (2) communicate the exact place of the activity in case the person does not remember it, (3) becomes a sign of recognition that leads to trust in the person who wears it and uses TIP, involving and making the users part of the same system. To do this, it uses the voice and a GPS. The portable device, designed to be carried in hand, on clothes, in the pocket, or in the bag, can take place in the center of the home device. Figure 1 portrays the final prototype of TIP and the setting with the user.

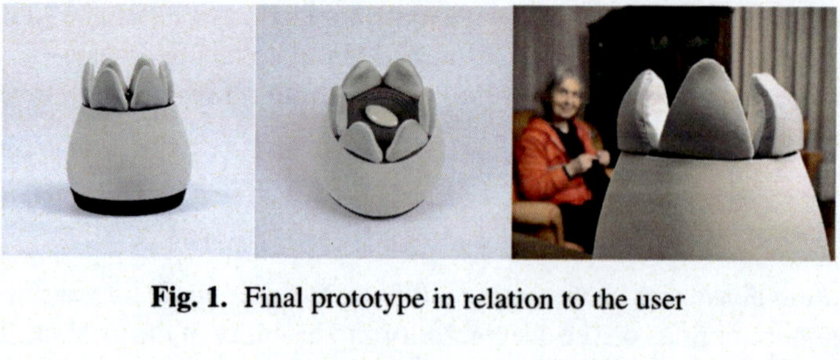

Fig. 1. Final prototype in relation to the user

4 Conclusion

TIP is a design project of a social robot for elderlies dealing with social isolation in which technology represents a means to improve the product as well as creating new human relationships. The research approach has allowed the designer to learn that collaboration with the end-user and healthcare figure is essential to build a correct problem setting. At the same time, the user realizes the usefulness of this product which helps him organize his day and meet new people while staying active and creating new social relationships. Tip has been designed for a private and self-sufficient user, but future research will improve and develop its use for nursing homes. The product could be enriched by allowing the robot to move autonomously on a surface. The portable device could become an object that allows a group of people to do activities or games together.

Acknowledgements. The project was developed by Giada Francescato, Ramiro Argañaraz, Francesco Argentino, Sara Ceccato, Giulia Patellaro and Gianmarco Poggiana. Authors would like to thank Roberta Pandolfo, the San Lorenzo Service Center and the group of elderly people. We would also thank Francesco Baldassarra.

References

1. Broekens, J., Heerink, M., Rosendal, H.: Assistive social robots in elderly care: a review. Gerontechnology **8**(2), 94–103 (2009)
2. Tennoe, T., Johannessen, A.: Technologies in care for older people. In: EPTA Report 2019 **5**(2), 90–100 (2019)
3. Istat Indicatori Demografici. http://www.istat.it. Accessed 25 Sept 2020
4. Courtin, E., Knapp, M.: Social isolation, loneliness and health in old age: a scoping review. Health Soc. Care Community **25**(3), 799–812 (2015)
5. Science Engineering Medicine. http://www.nap.edu. Accessed 20 Sept 2020
6. Krell-Roesh, J.: Quantity and quality of mental activities and the risk of incident mild cognitive impairment. Neurology **93**(6), 548–558 (2019)

Education in and with Inclusive Robots

Exploiting VR and AR Technologies in Education and Training to Inclusive Robotics

Maria Pozzi[1](✉), Unnikrishnan Radhakrishnan[2], Ana Rojo Agustí[3], Konstantinos Koumaditis[2], Francesco Chinello[2], Juan C. Moreno[3], and Monica Malvezzi[1]

[1] University of Siena, Siena, Italy
maria.pozzi@unisi.it
[2] Aarhus University, Aarhus, Denmark
[3] Spanish Research Council, Madrid, Spain

Abstract. Nowadays, robotics applications are no more limited to industrial rigid production cells or research laboratories and are spreading in several domains in which interaction with humans is required. In this context, educating professionals, workers, and students in the use and understanding of robotic systems is paramount to ensure the social acceptance and uptake of interactive robots. At the same time, the request for innovative training and teaching methodologies is growing, and virtual and augmented reality technologies as educational tools are gaining the attention of public institutions, companies, and healthcare facilities. In this paper, we focus on how VR/AR technologies can be exploited to teach robotics and to train in the use of specific robot platforms, providing a review of the available resources and discussing their advantages and disadvantages in the "inclusive robotics" perspective.

Keywords: Robotics education · Inclusive robotics · Virtual reality · Augmented reality

1 Introduction

Robots represented a step-change in industrial automation and are now a fundamental pillar of the so-called fourth industrial revolution [1]. Their presence in homes, hospitals, shops, and other service environments, however, is still limited. Several technical barriers still need to be tackled, including perception, power supply, safety, and human-robot interfaces, but also non-technical challenges are present. To facilitate the social uptake of interactive robots, interdisciplinary questions about the socio-economical, legal, and ethical impact of robotics on society need to be answered, and effective methods and resources to spread the knowledge of robotics-related topics must be developed [2]. Robots can play multiple roles in education: they can be the subject of the learning process [3, 4], or they can be used to teach other STEM subjects [5]. Recently, also virtual, mixed, and augmented reality technologies have become important tools for training students and workers [6–9].

M. Malvezzi et al. (Eds.): EDUROBOTICS 2021, SCI 982, pp. 115–126, 2021.
https://doi.org/10.1007/978-3-030-77022-8_11

The need for innovative training and education tools, allowing distance learning, possibly using virtual environments, has become evident during the lockdown period due to the COVID-19 pandemic in 2020 [10]. The closure of schools and universities in 192 countries has interrupted the education of nearly 1.6 billion students, representing 90% of the world's student population (UNESCO Institute for Statistics data[1]). In Europe, the closure of schools of all types and levels involved approximately 76.2 million students and 6.3 million teachers (Eurostat data[2]).

In this paper, we revise VR and AR resources for educating in robotics or in the use of specific robotic systems. After an overview of the advantages and disadvantages of using VR and AR for training and education (Sect. 2), we outline the challenges faced when teaching robotics at different levels and describe existing VR and AR tools applied to robotics (Sect. 3). In Sect. 4, we discuss future perspectives related to the use of innovative tools to teach and learn robotics.

2 VR/AR Technologies for Training and Education

Virtual Reality (VR) has been given a variety of definitions, most of the times dictated by market requirements and technology advancements. Milgram and Kishino [11], laid out the Reality-Virtuality continuum (see Fig. 1), in which the Virtual Environment is set at the completely opposite side with respect to the Real Environment (RE), since it provides an entirely virtual representation of the information. Between RE and VR, there are a variety of technological solutions aimed at "mixing" the image captured from the real scene with the digital content. Products proposed in the last ten years, including entertainment (e.g., PokemonGo) and communication (e.g., Snapchat) technologies increased the popularity of the term Augmented Reality (AR) to indicate platforms which overlay virtual images on real ones, but usually do not allow the interaction between the user and the virtual information in the scene. Mixed Reality (MR) is frequently referred to as an extension of AR, in which real and virtual elements can interact with one another, and the 3D content reacts to the user in the same way as it would do in the real world.

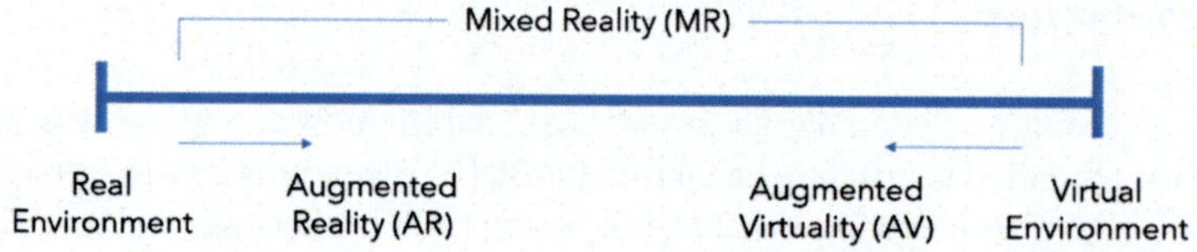

Fig. 1. Reality-Virtuality continuum [11]

Both VR and AR became very famous in the past decade thanks to advances in motion tracking, graphics processing power, and display technologies. Solutions became rapidly popular and relatively inexpensive, creating sets of tools including 6DOF-tracked hand controllers and headsets, the latter used as wearable immersive visors, rather than sensor

[1] https://unstats.un.org/unsd/ccsa/documents/covid19-report-ccsa.pdf.

[2] https://ec.europa.eu/eurostat/en/web/products-eurostat-news/-/DDN-20200604-1.

cameras. Device-oriented solutions that mix virtual objects to real scenarios, creating a hybridization of the original terms “virtual” and “real” into mixed reality, have also been proposed, including, e.g., Hololens (Microsoft Inc.).

Given this rich scenario, in which technological advancements found a florid market and fast development, relevant scientific results came to support the investigation of training and education with this technology. A key feature of VR is the “immersion” defined as the “experience of being immersed in virtual environments” [12]. Immersive VR (IVR) systems, in which devices with headsets can “isolate” the user from external visual cues, create a highly engaging experience. Isolation from visual cues and immersion, brought by IVR technology, are desirable features in the training domain [13–15].

Chittaro and Bottussi proposed an immersive serious game delivered through a head-mounted display for teaching aviation safety procedures and found several advantages with respect to conventional training techniques [16]. Krokos et al. compared IVR with non-immersive desktop VR condition for undergoing memory training and found that subjects showed better recall capabilities when immersive training was applied [17].

2.1 VR/AR for Professional Training

Given the promising results and the growing market interest, VR and IVR technologies are becoming more and more popular in industry and healthcare sectors where training environments are particularly challenging for safety, cost, or feasibility. In [18], for example, the authors proposed a solution based on VR to teach bimanual assembly skills to workers in factories. In the healthcare field, most of the previous works aim at teaching specific procedures to operators or at preparing them for the interaction with patients. Vaughan et al. [19], for example, created an IVR training system for paramedical operation to improve skills for needle cricothyroidotomy and chest draining, Butt et al. [20] used IVR and serious gaming to teach catheterization skills, whereas Shorey et al. [21] investigated the use of a desktop VR setup to train nurses in interaction with patients (Virtual Counseling Application).

Recently, a smartphone-based VR-application (OpenJustice) has been applied with promising results to legal education [22].

AR has been found effective for training professionals in different areas. In industrial scenarios, Webel et al. [23] introduced a framework based on the use of AR software and a tablet for teaching assembly and maintenance skills, enhancing the tasks comprehension, while Gavish et al. [24] found better effectiveness of AR with respect to VR in industrial assembly training for maintenance skills. Recently, Catal et al. [25] developed a game-based AR platform to train employees on building evacuation scenarios, whereas Rojas-Muñoz et al. [26] positively evaluated the use of AR for telementorization of practitioners in cricothyroidotomy procedures.

2.2 VR/AR for Students’ Education

It is indeed in the education sector that the cost-effectiveness and the technological advancements of VR technology are reflecting a great interest. Radianti et al. [13], in their survey paper, found that the most used immersive VR technologies in higher education are Head-Mounted Displays (HMDs), such as Oculus Rift (Facebook Technologies, LLC.) or HTC Vive (HTC Corporation).

Simeone et al. [27] explored a training scenario in which IVR was used to train college students and found that the presence of the instructor (in the virtual environment) has beneficial effects for the learning process. In the evaluation of Lifelique Museum, Allcoat and von Mühlenen [28] taught plant cells to college students with two different methods, one based on IVR, and one based on videos and textbooks. Participants who tested the IVR condition reported engagement, connected to a better learning experience, increased positive emotions and memory recall abilities. Many commercial VR headsets embed eye trackers and Rahman et al. [29] proposed to use this feature to record students' eye-gaze during lectures, thus allowing teachers to identify distracted or confused pupils and promptly guide their focus onto the important parts of the lesson.

Augmented reality is largely studied in educational contexts, too. Tobar et al. [30] used an AR game-based learning technique to promote reading comprehension among school students and compared it to traditional approaches. Results showed increased motivation in learning and better abilities in problem-solving and socialization. ARLIS, an instruction system created by Chen and Tsai [31], supported the investigation on learning performance with respect to traditional textbooks. Erbas and Demirer [32] investigated the use of AR-based activities in a ninth-grade biology course and found that these activities led to increased students' motivation and course success with respect to conventional methods. Sahin and Yilmaz [33] developed an AR booklet representing the Solar System resulting in an improvement of the motivation and learning achievements of students.

3 VR/AR Technologies for Robotics Training and Education

Table 1. Main outcomes of the literature review

VR/AR technologies for robotics training and education		
Training		Education
Industry	Healthcare	
Who: workers What: use of specialized robotic systems VR systems: Roldán et al. [34] Pérez et al. [35] Haruna et al. [36] AR/MR systems: Pai et al. [37]	Who: practitioners (surgeons, nurses, etc.), customers/patients What: use of specialized robotic systems, execution of robot-assisted procedures, robot aided rehabilitation exercises VR systems: Peral-Boiza et al. [38], Wang, et al. [39] Knopp et al. [40], Mariani et al. [41] Raison et al. [42], De la Iglesia et al. [43] Grimm et al. [44] AR/MR systems: Christensen et al. [45] Chowriappa et al. [46]	Who: students What: functioning, programming, and control of robots VR systems: Crespo et al. [47] Theofanidis et al. [48] Román-Ibáñez et al. [49] AR/MR systems: Jara et al. [50] Cheli et al. [51] Krajník et al. [52] Quintero et al. [53] Ostanin et al. [54]

The main outcomes of our literature review on robotics training and education through VR and AR technologies are summarized in Table 1 and discussed in the following sections. We decided to classify the selected papers based on the potential users of the described systems. Sect. 3.1 describes works addressing the training of professional operators in the use of specific robotic platforms or in the execution of robot-assisted procedures in two different main domains: industry and healthcare. Sect. 3.2, instead, analyses systems that are more suitable for training students and beginner robot programmers.

3.1 VR/AR-Based Training in the Use of Robots

The widely used computer-based training for robotics instruction now comprehends the benefits of extended-reality technologies to enhance the user experience. Current VR/AR solutions are aimed at covering the two fundamental needs of *i)* training in the use of specialized robotic systems, and *ii)* training in robot-assisted procedures, differentiating those intended for professionals or for customers/patients. These technologies have been applied in industrial scenarios as well as in the training of health professionals.

Technological, industrial examples include virtual environments intended for training system operators, allowing them to program and check whether a certain robotic system will do what is expected. In [35], a VR framework to train operators in the use of an industrial robot is presented. The adoption of VR ensures workers' safety and the immersiveness of the proposed solution improves training effectiveness. Different potential users (robotic engineers, robot operators, and assistant operators) evaluated the system and found it usable and useful. Haruna et al. [36] developed a VR-based system that visualizes haptic information through perceptual images overlaid at the contact points of a remote robotic hand. They showed that such a "visual haptics system" can help the pre-training of operators that need to learn how to control a robotic manipulation system from remote. With the aim of reducing lead time and lowering manufacturing costs, Pai et al. [37] evaluated the design of an augmented reality interface aimed at strengthening user's understanding and at improving interaction with the manufacturing environment. The AR interface guides users from the layout planning phase to the prototyping of the product of a fully automated work cell.

Other interesting approaches are those virtual training platforms that integrate data mining or predictive algorithms able to perform operator functions and support their decision making. Both strategies foster the adaptation process, understanding, coordination, and workload of trainees. The study by Roldán et al. [34] focused on the creation of an immersive interface system which enables multi-robot interfaces training while implementing a layer for evaluation/prediction of operators capabilities: workload, situational awareness, stress, and trust.

Virtual simulators are extremely valuable for health professionals who work with multiple robots in different scenarios. Acknowledging that realistic imitation of robotic surgery has allowed safer training for surgeons and patients rather than by caseload practices, surgeons have adopted these technologies to improve their technical proficiency, mostly with wristed instruments used in laparoscopy and endoscopy procedures. Peral-Boiza et al. [38] reported on the suitable use of their virtual reality training platform for robot-assisted flexible ureteroscopy interventions which enables real-time user interactions in a wide range of urolithiasis scenarios. Likewise, Wang et al. [39] concluded in their study that urologists improved their skills for vesicourethral anastomosis and shortened the learning curve when using virtual training for anastomosis. Knopp et al. [40] created a robotic immersive VR surgical training scenario to teach trainees hip replacements and, recently, Mariani et al. [41] showed the effectiveness of VR for robotic surgical adaptive training.

The continuous and growing generation of VR solutions for surgical training forces the industry to settle on a standard. It is worthwhile to obtain specific metrics for all generic sequential tasks across different robotic skills exercises to ensure that competencies in robot-assisted surgery have been achieved. For this reason, several research groups from medical institutions have introduced a benchmark score for virtual robotic simulations in order to determine a competency-based virtual robotic training curriculum [42].

Also, AR-based solutions have been proposed for the training of surgeons, with the intent of facilitating the communication between trainer and trainee in minimally invasive surgery [45] or in specific operations. Chowriappa et al. [46], for example, showed that using an AR environment boosts robot-assisted surgery skills acquisition for urethrovesical anastomosis with minimal cognitive load.

On the other side, solutions based on VR, AR, and customizable games have been proposed to improve the user interface of robotic equipment for patient rehabilitation, aiming at increasing the interest of patients so that they keep performing their exercises [55, 56]. Exoskeletons connected to VR systems may allow a patient to perform personalized exercises with immersion in a motivational environment. The approach by De la Iglesia et al. [43] consisted of a context-aware VR system focused on patient follow-up on elbow rehabilitation. This solution moves towards future telerehabilitation offering a low-cost exoskeleton combined with different medical sensors to capture relevant patient data enabling remote medical monitoring, cloud rehabilitation exercises, and cloud storing data. Taking virtual reality beyond a simple motivational training space, but generating an environment fed by parameters captured by sensors and robotic systems, Grimm et al. [44] proposed an ambitious goal of automating the treatment for recovery of upper limb movements post-stroke. Their approach targeted the improvement of upper limb range of motion based on the adaptation of the virtual environment to the patient's robotic-assistance dependence during unsupervised adaptive training of reach-to-grasp exercises.

3.2 Teaching Robotics and with Robotics Through VR/AR

Robotics is a learning subject requiring multidisciplinary knowledge and skills: using robots in education activities engages students and improves their technical and non-technical capabilities, including problem-solving, analytical and critical thinking, reflection, and creativity [57, 58]. Robots are also a useful tool for promoting cross-subject projects, supporting, for example, the learning of other STEM disciplines (physics, biology, etc.). Teaching activities involving robots are becoming increasingly common both in schools and universities, where robots are often used in engineering and computer programming courses.

In the previous sections, we have pointed out that VR and AR technologies are becoming popular tools to enhance education and training at different levels. They can be exploited to teach professionals and end-users how to use a certain robotic system, allowing mitigation of the costs due to the physical building of the system and the possible damages provoked by inexperienced users.

In the context of robotics education, VR can be used to create training interfaces and virtual laboratories [59], while AR technologies can enrich the students' direct experience with the robots. Here we analyze works in which VR/AR/MR are used to teach robotics-related topics or to ease robot programming.

Several previous works proposed the use of VR to teach students and operators how to program robot manipulators [47, 48]. These works usually rely on a simulator of the robot and provide the user with an immersive interface to command it. Crespo et al. [47] showed that VR based training is effective for engineering students. Theofanidis et al. [48] showed that a VR-based solution allows users to control a robot arm in a more efficient way with respect to a less realistic interface. VR, however, is itself outperformed by the kinaesthetic teaching, in which the user physically interacts with the robot.

Román-Ibáñez et al. [49], focused on undergraduate students' education and introduced a low-cost IVR environment to teach how to program robotic manipulators. The use of a virtual laboratory avoids the need for performing experiments in real conditions, which usually costs time, money, and energy. Even though a similar result could be achieved using a simulator (e.g., the Robotics Toolbox [60]), immersive VR can foster students' engagement.

AR/MR technologies, likewise robots, can be used to make concrete, visible, and even touchable otherwise abstract and intangible concepts. One of the first works combining AR and robotics education describes an AR-based interface allowing students to simulate and teleoperate a robotic arm [50]. AR-based activities can also help students to see what happens behind the scenes, inside the robot, for example reading sensor values or visualizing the code that is currently executing [51], or visualizing the state of the robot [52]. Cheli et al. [51] underlined that using AR based interfaces can help students to debug their code and discuss together how to fix problems. However, the authors also recognized that AR systems could have usability issues and need to be properly designed to be successful.

MR can be used to intuitively program manipulators by superimposing a holographic robot over the real one and allowing the visualization and modification of the robot trajectory [53, 54]. Quintero et al. [53] proposed an AR system thanks to which the user could interact with a robot arm through speech and gestures. This solution allowed to program the operations of the manipulator more efficiently than using kinaesthetic teaching. The observed differences were mainly due to the fact that the use of the AR interface required fewer human motions.

While in [47, 49–52] authors aimed at creating educational and training platforms for students of different levels, in [48, 53, 54], authors presented systems that can ease robot programming thanks to the use of VR/AR/MR technology. Results obtained by the latter could inform the development of future educational platforms.

The technologies used in the works described in this section are summarized in Table 2. In most of the cases, HMDs are employed and there is a focus on robot manipulators.

Table 2. Technologies employed in the works analyzed in this section

Paper	AR/VR system	Robot
Crespo et al. [47]	Oculus Rift (Facebook Technologies, LLC.), Razer Hydra Joysticks (Razer Inc.)	Mitsubishi RV-M1 Robot Manipulator
Theofanidis et al. [48]	Oculus Rift (Facebook Technologies, LLC.), Leap Motion (Ultraleap)	4-DOF Barrett Whole Arm Manipulator (WAM) (Barrett Technology)
Román-Ibáñez et al. [49]	Cardboard VR glasses with a smartphone placed inside	Industrial Robot Manipulator
Jara et al. [50]	3D visualization on a screen of real information from the robot, complemented with some virtually generated data.	Scorbot ER-IX Robot Manipulator (Eshed Robotec Inc.)
Cheli et al. [51]	iPad (Apple Inc.) with Thingworx View (PTC)	EV3 Mobile Robot (LEGO® MINDSTORMS® kit)
Krajník et al. [52]	Robot information overlaid on the real video	AR-Drone Quadcopter
Quintero et al. [53]	Hololens (Microsoft Inc.)	7-DOF Barrett WAM (Barrett Technology)
Ostanin et al. [54]	Hololens (Microsoft Inc.)	UR10 (Universal Robots) and LBR iiwa (KUKA AG) Robot Manipulators

Notwithstanding the optimism and enthusiasm arising from the described promising experiences, there are still some remaining challenges to the successful exploitation of VR and AR in robotics teaching, especially in the first levels of education, in which educational robotics is not yet fully structured in educational programs and is not homogeneously distributed over different countries and regions. In addition, teachers often lack the knowledge of robotics in their own schooling or training courses. The introduction of educational robotics and the effective use of AR and VR technologies in this context requires reviewing existing pedagogical approaches for many teachers. Therefore, effective professional development of teachers is a key aspect.

4 Conclusions and Perspectives

Technology constitutes an important resource for many aspects of our lives, including training and education. In particular, Virtual and Augmented Reality applications are living a fruitful development phase in terms of accessible devices and resources, and effective applications. At the same time, robots, initially confined in high-tech research centres and in large companies with automated production lines, are nowadays becoming familiar to the whole society.

The need for adequate and accessible tools for learning robotics and training with robots at different education levels and for different purposes is becoming significant. From the pedagogical point of view, several experiences demonstrated that robots are an effective tool to enable people with different forms and levels of expertise to come together and express their ideas, demonstrate problems, construct shared knowledge and communicate potential solutions to others. In fact, robots can give concreteness to otherwise abstract concepts and issues.

In this paper, we discussed how technologies leveraging virtual reality or different levels of mixed reality can be used in robotics training and education. We reviewed the applications that, to the best of our knowledge, are most relevant, by highlighting the positive effects and the potentialities of such systems.

The use of VR interfaces allows teachers to deliver laboratory lectures without the need for large and expensive infrastructures. At the same time, avoiding the direct interaction with robots ensures a safe learning environment. VR based educational activities cannot completely replace the training with real robots, but they can serve as a preliminary learning step before programming and using real devices.

AR and MR interfaces do not substitute the interaction with the real robot, but rather they enrich it, giving the user intuitive tools to program the robot and monitor its state. Applications in robotics education are still limited as the use of these technologies requires additional training for students, who must learn to manage the interface as well as the robot.

Other drawbacks connected to the use of VR and AR for robotics education are mainly related to the accessibility of these resources, in terms of costs and educators' training. It is worth mentioning that as soon as a system gets out of a specific applicative environment, and starts being adopted by the society as a whole, the technological, scientific and engineering aspects become only a part of the problem, as also human-related and non-technical aspects, such as psychology, law and ethics must be considered. We, therefore, believe that an inclusive approach to robotics and a harmonic dialogue between technical and non-technical experts is fundamental for the fruitful exploitation of VR, AR, and robot technologies in education.

Author Contributions. All authors equally contributed to the paper.

Funding. This work was supported by the EU Horizon 2020 research and innovation programme under GA No 780073 INBOTS (Inclusive Robotics for a Better Society).

References

1. Schwab, K.: The fourth industrial revolution. Crown (2017)
2. Alimisis, D., Loukatos, D., Zoulias, E., Alimisi, R.: The role of education for the social uptake of robotics: the case of the eCraft2Learn project. In: Inclusive Robotics for a Better Society, Cham, pp. 180–187 (2020). https://doi.org/10.1007/978-3-030-24074-5_30

3. Pozzi, M., Prattichizzo, D., Malvezzi, M.: On-line educational resources on robotics: a review. In: Inclusive Robotics for a Better Society, Cham, pp. 141–147 (2020). https://doi.org/10.1007/978-3-030-24074-5_25
4. Pozzi, M., Malvezzi, M., Prattichizzo, D.: MOOC on the art of grasping and manipulation in robotics: design choices and lessons learned. In: Robotics in Education, Cham, pp. 71–78 (2019). https://doi.org/10.1007/978-3-319-97085-1_7
5. Ferrarelli, P., et al.: Improving students' concepts about newtonian mechanics using mobile robots. In: Lepuschitz, W., Merdan, M., Koppensteiner, G., Balogh, R., Obdržálek, D. (eds.) Robotics in Education, vol. 829, pp. 113–124. Springer International Publishing, Cham (2019)
6. Martín-Gutiérrez, J., Mora, C.E., Añorbe-Díaz, B., González-Marrero, A.: Virtual technologies trends in education. Eurasia J Math Sci T **13**(2), January 2017. https://doi.org/10.12973/eurasia.2017.00626a
7. Ibáñez, M.-B., Delgado-Kloos, C.: Augmented reality for STEM learning: a systematic review. Comput. Educ. **123**, 109–123 (2018). https://doi.org/10.1016/j.compedu.2018.05.002
8. Bellalouna, F.: New approach for industrial training using virtual reality technology. Procedia CIRP **93**, 262–267 (2020). https://doi.org/10.1016/j.procir.2020.03.008
9. Chinello, F., Koumaditis, K.: Virtual immersive educational systems: early results and lessons learned. In: SIGGRAPH Asia 2019 Posters, Brisbane QLD Australia, pp. 1–2, November 2019. https://doi.org/10.1145/3355056.3364586
10. Pozzi, M., Prattichizzo, D., Malvezzi, M.: Accessible educational resources for teaching and learning robotics. Robotics **10**(1), 38 (2021). https://doi.org/10.3390/robotics10010038
11. Milgram, P., Kishino, F.: A taxonomy of mixed reality visual displays. IEICE Trans. Inf. Syst. **77**(12), 1321–1329 (1994)
12. Suh, A., Prophet, J.: The state of immersive technology research: a literature analysis. Comput. Hum. Behav. **86**, 77–90 (2018)
13. Radianti, J., Majchrzak, T.A., Fromm, J., Wohlgenannt, I.: A systematic review of immersive virtual reality applications for higher education: design elements, lessons learned, and research agenda. Comput. Educ. **147**, (2020)
14. Koumaditis, K., Chinello, F., Mitkidis, P., Karg, S.: Effectiveness of virtual versus physical training: the case of assembly tasks, trainer's verbal assistance, and task complexity. IEEE Comput. Grap. Appl. **40**(5), 41–56 (2020). https://doi.org/10.1109/MCG.2020.3006330
15. Koumaditis, K., Chinello, F., Venckute, S.: Design of a virtual reality and haptic setup linking arousals to training scenarios: a preliminary stage. In: 2018 IEEE Conference on Virtual Reality and 3D User Interfaces (VR), Reutlingen, pp. 1–2, March 2018. https://doi.org/10.1109/vr.2018.8446528
16. Chittaro, L., Buttussi, F.: Assessing knowledge retention of an immersive serious game vs. a traditional education method in aviation safety. IEEE Trans. Vis. Comput. Graph. **21**(4), 529–538 (2015)
17. Krokos, E., Plaisant, C., Varshney, A.: Virtual memory palaces: immersion aids recall. Virtual Reality **23**(1), 1–15 (2019)
18. Carlson, P., Peters, A., Gilbert, S.B., Vance, J.M., Luse, A.: Virtual training: learning transfer of assembly tasks. IEEE Trans. Vis. Comput. Graph. **21**(6), 770–782 (2015)
19. Vaughan, N., John, N., Rees, N.: ParaVR: paramedic virtual reality training simulator. In: 2019 International Conference on Cyberworlds (CW), pp. 21–24 (2019)
20. Butt, A.L., Kardong-Edgren, S., Ellertson, A.: Using game-based virtual reality with haptics for skill acquisition. Clin. Simul. Nursing **16**, 25–32 (2018)
21. Shorey, S., Ang, E., Ng, E.D., Yap, J., Lau, L.S.T., Chui, C.K.: Communication skills training using virtual reality: a descriptive qualitative study. Nurse Educ. Today **94**, (2020)
22. McFaul, H., FitzGerald, E.: A realist evaluation of student use of a virtual reality smartphone application in undergraduate legal education. Br. J. Educ. Technol. **51**(2), 572–589 (2020)

23. Webel, S., Bockholt, U., Engelke, T., Gavish, N., Olbrich, M., Preusche, C.: An augmented reality training platform for assembly and maintenance skills. Robot. Auton. Syst. **61**(4), 398–403 (2013)
24. Gavish, N., et al.: Evaluating virtual reality and augmented reality training for industrial maintenance and assembly tasks. Interact. Learn. Environ. **23**(6), 778–798 (2015)
25. Catal, C., Akbulut, A., Tunali, B., Ulug, E., Ozturk, E.: Evaluation of augmented reality technology for the design of an evacuation training game. Virtual Reality, 1–10 (2019)
26. Rojas-Muñoz, E., et al.: Evaluation of an augmented reality platform for austere surgical telementoring: a randomized controlled crossover study in cricothyroidotomies. NPJ Digital Med. **3**(1), 1–9 (2020)
27. Simeone, A.L., Speicher, M., Molnar, A., Wilde, A., Daiber, F.: LIVE: the human role in learning in immersive virtual environments. In: Symposium on Spatial User Interaction, pp. 1–11 (2019)
28. Allcoat, D., von Mühlenen, A.: Learning in virtual reality: effects on performance, emotion and engagement. Res. Learn. Technol. **26** (2018)
29. Rahman, Y., Asish, S.M., Fisher, N.P., Bruce, E.C., Kulshreshth, A.K., Borst, C.W.: Exploring eye gaze visualization techniques for identifying distracted students in educational VR. In: 2020 IEEE Conference on Virtual Reality and 3D User Interfaces (VR), pp. 868–877 (2020)
30. Tobar-Muñoz, H., Baldiris, S., Fabregat, R.: Augmented reality game-based learning: enriching students' experience during reading comprehension activities. J. Educ. Comput. Res. **55**(7), 901–936 (2017)
31. Chen, C.-M., Tsai, Y.-N.: Interactive augmented reality system for enhancing library instruction in elementary schools. Comput. Educ. **59**(2), 638–652 (2012)
32. Erbas, C., Demirer, V.: The effects of augmented reality on students' academic achievement and motivation in a biology course. J. Comput. Assist. Learn. **35**(3), 450–458 (2019)
33. Sahin, D., Yilmaz, R.M.: The effect of augmented reality technology on middle school students' achievements and attitudes towards science education. Comput. Educ. **144**, (2020)
34. Roldán, J.J., Peña-Tapia, E., Martín-Barrio, A., Olivares-Méndez, M.A., Del Cerro, J., Barrientos, A.: Multi-robot interfaces and operator situational awareness: study of the impact of immersion and prediction. Sensors **17**(8), 1720 (2017)
35. Pérez, L., Diez, E., Usamentiaga, R., García, D.F.: Industrial robot control and operator training using virtual reality interfaces. Comput. Ind. **109**, 114–120 (2019)
36. Haruna, M., Ogino, M., Koike-Akino, T.: Proposal and evaluation of visual haptics for manipulation of remote machine system. Front. Robot. AI **7**, (2020). https://doi.org/10.3389/frobt
37. Pai, Y.S., Yap, H.J., Dawal, S.Z.M., Ramesh, S., Phoon, S.Y.: Virtual planning, control, and machining for a modular-based automated factory operation in an augmented reality environment. Sci. Rep. **6**, 27380 (2016)
38. Peral-Boiza, M., Gomez-Fernandez, T., Sanchez-Gonzalez, P., Rodriguez-Vila, B., Gómez, E.J., Gutiérrez, Á.: Position based model of a flexible ureterorenoscope in a virtual reality training platform for a minimally invasive surgical robot. IEEE Access **7**, 177414–177426 (2019)
39. Wang, F., et al.: The application of virtual reality training for anastomosis during robot-assisted radical prostatectomy. Asian J. Urol. (2019)
40. Knopp, S., Lorenz, M., Pelliccia, L., Klimant, P.: Using industrial robots as haptic devices for VR-training. In: 2018 IEEE Conference on Virtual Reality and 3D User Interfaces (VR), pp. 607–608 (2018)
41. Mariani, A., Pellegrini, E., Enayati, N., Kazanzides, P., Vidotto, M., De Momi, E.: Design and evaluation of a performance-based adaptive curriculum for robotic surgical training: a pilot study. In: 2018 40th Annual International Conference of the IEEE Engineering in Medicine and Biology Society (EMBC), pp. 2162–2165 (2018)

42. Raison, N., et al.: Competency based training in robotic surgery: benchmark scores for virtual reality robotic simulation. BJU Int. **119**(5), 804–811 (2017)
43. de la Iglesia, D.H., Mendes, A.S., González, G.V., Jiménez-Bravo, D.M., de Paz Santana, J.F.: Connected elbow exoskeleton system for rehabilitation training based on virtual reality and context-aware. Sensors, **20**(3), 858 (2020)
44. Grimm, F., Naros, G., Gharabaghi, A.: Closed-loop task difficulty adaptation during virtual reality reach-to-grasp training assisted with an exoskeleton for stroke rehabilitation. Front. Neurosci. **10**, 518 (2016)
45. Christensen, N.H., et al.: Depth cues in augmented reality for training of robot-assisted minimally invasive surgery. In: Proceedings of the 21st International Academic Mindtrek Conference, pp. 120–126 (2017)
46. Chowriappa, A., et al.: Augmented-reality-based skills training for robot-assisted urethrovesical anastomosis: a multi-institutional randomised controlled trial. BJU Int. **115**(2), 336–345 (2015)
47. Crespo, R., García, R., Quiroz, S.: Virtual reality application for simulation and off-line programming of the mitsubishi movemaster RV-M1 robot integrated with the oculus rift to improve students training. Procedia Comput. Sci. **75**, 107–112 (2015)
48. Theofanidis, M., Sayed, S.I., Lioulemes, A., Makedon, F.: Varm: using virtual reality to program robotic manipulators. In: Proceedings of the 10th International Conference on PErvasive Technologies Related to Assistive Environments, pp. 215–221 (2017)
49. Román-Ibáñez, V., Pujol-López, F.A., Mora-Mora, H., Pertegal-Felices, M.L., Jimeno-Morenilla, A.: A low-cost immersive virtual reality system for teaching robotic manipulators programming. Sustainability **10**(4), 1102 (2018)
50. Jara, C.A., Candelas-Herías, F.A., Fernández, M., Torres, F.: An augmented reality interface for training robotics through the web (2009)
51. Cheli, M., Sinapov, J., Danahy, E.E., Rogers, C.: Towards an augmented reality framework for k-12 robotics education (2018)
52. Krajník, T., Vonásek, V., Fišer, D., Faigl, J.: AR-drone as a platform for robotic research and education. In: International Conference on Research and Education in Robotics, pp. 172–186 (2011)
53. Quintero, C.P., Li, S., Pan, M.K., Chan, W.P., Van der Loos, H.M., Croft, E.: Robot programming through augmented trajectories in augmented reality. In: 2018 IEEE/RSJ International Conference on Intelligent Robots and Systems (IROS), pp. 1838–1844 (2018)
54. Ostanin, M., Mikhel, S., Evlampiev, A., Skvortsova, V., Klimchik, A.: Human-robot interaction for robotic manipulator programming in Mixed Reality. In: 2020 IEEE International Conference on Robotics and Automation (ICRA), pp. 2805–2811 (2020)
55. Asín-Prieto, G., et al.: Haptic adaptive feedback to promote motor learning with a robotic ankle exoskeleton integrated with a video game. Front. Bioeng. Biotechnol. **8**, 113 (2020)
56. Mubin, O., Alnajjar, F., Jishtu, N., Alsinglawi, B., Al Mahmud, A.: Exoskeletons With virtual reality, augmented reality, and gamification for stroke patients' rehabilitation: systematic review. JMIR Rehabil. Assistive Technol. **6**(2), e12010 (2019)
57. Nagai, K.: Learning while doing: practical robotics education. IEEE Robot. Autom. Mag. **8**(2), 39–43 (2001)
58. Alimisis, D.: Educational robotics: open questions and new challenges. Themes Sci. Technol. Educ. **6**(1), 63–71 (2013)
59. Potkonjak, V., et al.: Virtual laboratories for education in science, technology, and engineering: a review. Comput. Educ. **95**, 309–327 (2016)
60. Corke, P.: Robotics, vision and control: fundamental algorithms in MATLAB® second, completely revised, vol. 118. Springer (2017)

Educational Robotics Curricula: Current Trends and Shortcomings

Theodosios Sapounidis[1,2](✉) and Dimitris Alimisis[1]

[1] European Lab for Educational Technology-EDUMOTIVA, Sparta, Greece
alimisis@edumotiva.eu
[2] Department of Information and Electronic Engineering, International Hellenic University (I.H.U.), Thessaloniki, Greece
sapounidis@ihu.gr

Abstract. Nowadays, Educational Robotics (ER) is flourishing at research and didactic level aiming to enhance higher-level thinking and skills. Families, schools, and educational institutes try to offer ER activities for children by utilizing various existing technologies and curricula. Despite the materials that have been created and the available technologies and curricula, it seems that not enough attention has been paid to the comprehensiveness and homogeneity of the existing curricula. The paper introduces the importance of features usually ignored in the available curricula such as icebreakers, collaboration scripts, level of learning guidance, and provision of multilingual support. Then, the paper presents a review of indicative curricula, covering many well-known robotic technologies and systems to extract common elements and simultaneously highlight shortcomings of the current approaches. Finally, the paper argues about the need for a paradigm shift in ER curricula that will incorporate the maker and Do-It-Yourself (DIY) culture.

Keywords: Educational robotics · Curricula development · Curricula review

1 Introduction

Educational robotics is a learning tool that allows students to build and then program a robotic entity, often using a simple programming language. The utilization of these technologies has its roots in Piaget's ideas on constructivism and in Seymour Papert's work started with the Logo programming language [1]. Learning with these systems is stimulated through collaboration and interaction with real-world problems, cultivating creativity, and problem-solving abilities with direct feedback and observation of the results of a possible solution [2, 3]. A significant educational characteristic in ER is that knowledge is formed mostly through play in small groups. This setting supports students' reasoning about the choices they make to overcome problem-solving challenges [4, 5].

To design and program a robotic artifact to make even a simple task can be extensively demanding for students. This is not only because the students have to understand how the robot works but also for the needed effort to follow the development and programming instructions which usually are contained in the related curricula. The variety

M. Malvezzi et al. (Eds.): EDUROBOTICS 2021, SCI 982, pp. 127–138, 2021.
https://doi.org/10.1007/978-3-030-77022-8_12

of the commercial robotic systems (e.g. [6–8]) along with the heterogeneity of the existing programming languages (e.g. [9]) is also reflected in the diversity of the offered curricula. ER curricula are offered by multiple developers which are not always reliable and rarely use common principles and practices. Therefore, this paper initially makes an introduction to some characteristics that are important for an ER curriculum but seldom considered. Then, a review of indicative curricula is presented covering some well-known robotic systems and aiming to extract common characteristics and concurrently highlight shortcomings of the current approaches. Finally, the paper argues about a) the need for a paradigm shift that will include the maker and Do-It-Yourself (DIY) movement and b) the need for the development of a generic curriculum model that might serve as a guide for developers while at the same time will allow freedom of choices according to target groups and specific objectives.

2 ER Curricula: Some Important Characteristics

The curriculum is usually the formulation of a teaching proposal and plan. A curriculum may refer to the teaching of an entire educational level or class in a particular subject or a set of subjects. Typical parameters of a curriculum are its objectives, the method it uses, the procedures it recommends, its evaluation proposals, time schedule and the course content. However, each curriculum developer can organize a lesson plan freely depending on the domain of usage, the target group or other factors. Obviously, the choices that will be made in relation to the elements that a curriculum will contain, or not contain, will determine its success and consequently the knowledge and skills that learners will acquire. The literature and recent research have shown that there are elements that might improve the competence of the curriculum to help learners effectively acquire new knowledge and skills. Such elements include, among others, the icebreakers, collaboration scripts, level of learning guidance and multilingual content which are rarely included in the educational robotics curricula [6].

2.1 Relieving the Learning Environment with Icebreakers

Many teachers and instructors have faced situations where one or more hesitant learner refuses to leave his or her protective shell and participate in a class. These learners need help to feel more open, to relieve inhibitions, to begin conversations, and get acquainted with the other group peers [10]. To promote the needed safe environment teachers can organise simple activities (icebreakers) to assist learners simply learn students' names, cultivate curiosity and positive attitude for a subject, bring humor into the course and so build the appropriate mood in the class [11]. Ice-breaking activities are usually introductory events that might contribute at the beginning of a course, to stimulate student's attention and curiosity about a subject, to assist team bonding, and to help participants feel comfortable [11]. These activities are usually short and may involve, writing (e.g., some personal information), talking, singing, or have a physical action (dancing, jumping). They intend to bridge the beginning of an event and let teachers connect and better understand students' needs and backgrounds and so build students' positive attitudes for the learning to follow [10]. These kinds of activities are applied to a variety of group settings, formal and informal learning, different age groups, and many domains (e.g., learning, meetings, etc.).

Although, there is a lack of systematic work that provides empirical evidence on the advantages of using ice-breaking activities at the beginning of a class some preliminary studies exist [12]. In detail, studies have shown that icebreakers may increase student's attention and help instructors to easier link students to the class processes [11]. Other, studies have shown that audiovisual materials used in ice-breaking activities may have positive effects on motivation and participation (e.g. [13]) and might prepare the students for deeper collaborative learning [14]. Simultaneously the combination of icebreakers and re-energizers in class may help to improve student participation and so enhance learning [10]. Unfortunately, in the domain of educational robotics, there is not systematic integration of icebreakers in curricula and this might be a restrictive factor for both students and educators.

2.2 Collaboration Scripts in Robotics Projects

Collaboration scripts aim to improve the collaboration between group members and act as a guideline on how a group of learners have to solve collaboration problems. [6]. To achieve this, collaboration scripts provide detailed strategies and a structure for the group members' interaction [15, 16]. Studies have shown that by scripting the pear collaboration procedures might reinforce argumentation skills and improve peer review and tutoring [16, 17]. Simultaneously, other studies have shown that if students left without any collaboration support might fail to engage in futile collaboration affecting their learning and performance [2]. This is why researchers and teachers strongly recommend the use of collaboration scripts in learning activities [17].

Focusing on educational robotics, one of the ER benefits appears to be the opportunities offered for social interaction and collaborative learning mostly because group members can easily share ideas, experiences, and roles [6]. This is because peers with no difficulty can monitor other team members' work gaining access to knowledge and skills. Even though, many ER researchers have highlighted the importance of students' collaboration mostly for cognitive, metacognitive, and social reasons we can hardly find collaboration scripts in the related curricula. The inclusion of such materials in ER curricula might have a significant effect on student's collaboration skills, engagement, active participation, and possibly learning (e.g. [18, 19]).

2.3 Guidance in Educational Robotics Curricula

The importance of guidance during learning has attracted the attention of researchers a long time ago. The theoretical background of the introduction of learning guidance is related with human working memory capacity which is limited and therefore, poor or no guidance, might overload students' working memory limiting learning speed and abilities. Many researchers claim that guided learning is fully compatible with constructivism model which promotes knowledge construction through exploration and discovery on real word stimulating problems [20]. Thus, it is supported that learning should be supervised and guided through several strategies (e.g. [20–22]). At the same time, other researchers argue that minimum (or even none) guidance might seem attractive but is less effective than strong guidance, especially for beginners. [23]. Thus, Clark et al. [24], proposes the combination of direct corrective feedback with strong guidance, at each step of the learning procedure, as a way to enhance learning. On the contrary, other researchers claim that strong guidance might cause workload for both teachers and students [25] and perhaps is completely useless when students obtain a significant knowledge level in the under-study field [26, 27]. Based on the above, it is clear that the main debate about guidance is about the level and form of guidance along with the strategies to gradually "fade out" guidance [28, 29]. Moreover, we can assume that for successful guidance, educators and curricula developers should bear in mind the human cognitive architecture and should avoid to cognitively overload students. Also, successful guidance must consider previous knowledge of the under- study field (novice or expert) along with the difficulty of the new knowledge domain.

Focusing on educational robotics, we know that when students start learning robotics they need substantial support and assistance before they make progress with their projects [30]. Particularly for novice learners it is essential to be introduced with care. Exposure to excessive levels of frustration should be avoided in order to avoid disappointment among learners [30]. Although, failure is an important part of learning and trying several different solutions (trial and error) is a completely acceptable way of learning, guidance might be the border between effective learning and permanent failure and discouragement. Although the value of guidance in learning and especially in robotics seems to be beyond doubt, there is no corresponding utilisation in educational robotics curricula. This leaves teachers and students without substantial support for learning guidance. Consequently, when teachers use a robotics curriculum, they are not aware how to help the children, nor the points where this help is necessary.

2.4 Multilingual Content

Regarding the educational material and the languages it has been developed, many researchers have found that the failure to utilise pupils' language and culture might have a negative impact on their motivation and achievement [31]. In contrast, addressing cultural and linguistic students' diversity as a resource rather than as a discrepancy, by adopting multilingual practices, can promote performance and communicative competence [32]. Although the need but also the value of multilingualism in curricula has been emphasized (e.g. [33]), in the field of educational robotics multilingual curricula are rather rare. It is characteristic that there are cases where educational material for children has been developed in a language that children most probably do not know as long as they are not native speakers. Therefore, we propose to exclusively develop multilingual curricula that can be used more effectively by the international audience (teachers and students) interested to engage in educational robotics.

3 Indicative Curricula

To identify representative curricula in educational robotics, we did a manual search on Google. We focused on curricula that cover different technologies and systems in order to have a broader view of the existing situation. Obviously, this paper does not present all the available curricula that can be found on the internet. For the needs of this study, only some indicative examples of the most representative curricula were selected and are presented below.

3.1 WeDo2 by Lego[1]

It contains multilingual teacher guides and preparation materials for WeDo 2.0. The material was prepared by Lego and is dedicated to teachers and primary school students. There are three project types, (a) getting started projects for novice (b) guided projects, which are related to specific curriculum ideals, and contain step-by-step instructions (c) open projects, which provide a more open-ended experience. It proposes an indicative time schedule and four phases for the project completion (explore, create, test, and share). It also covers theoretical issues like computational thinking and provides teachers with material for building and programming along with discussion questions/answers, assessment support, etc. The material is presented in pdfs, videos, webpages, and is free of charge.

3.2 Micro:Bit by Micro:Bit Educational Foundation[2]

The curriculum is mostly based on webpages along with videos. The instructions are divided into steps and there are separate materials for students and teachers. There are also items offered for free in pdfs, word, presentations, and source code files. Particularly for teachers, there are notes with extra information, code examples, and possible short extensions. Finally, in the curriculum, there are possible learning outcomes along with a time schedule estimation.

3.3 Multiple Systems by Sparkfun[3,4]

Materials for many platforms (Arduino, micro:bit, Raspberry Pi, Paper Circuits) and programming languages developed by Sparkfun which is an online retailer, particularly active in open source tech. The material is presented in many forms (pdfs, videos, webpages, ppts, etc.) and follows a dissimilar philosophy for the different technologies. There are learning objectives, time schedules and calendars, activities, open projects, and examples for the project expansion along with assessments. Usually, the material can be separated for students and teachers and in some cases, there are extra lecturing slides for the teacher.

1 https://education.lego.com/en-au/support/wedo-2/teacher-guides.

2 https://microbit.org/en/2017-03-07-javascript-block-resources/#lessons_a.

3 https://www.sparkfuneducation.com/curriculum.html.

4 https://learn.sparkfun.com/resources/39?_ga=1.93270749.1176615929.1473301234.

3.4 Multiple Systems by Carnegie Mellon Robotics Academy[5,6,7]

It is a series of curricula developed by Carnegie Mellon's Robotics Academy. The curricula cover many robotic platforms (VEX IQ, VEX CORTEX, VEX V5, Lego EV3, Parallax BOT) and use pdfs, videos, and webpages. Parts of the content are free while most of them not. There are tips for preparing the class, optional activities, writing reflection questions, worksheets, rubrics, additional handouts. It also covers some theoretical issues related to educational robotics and programming. The material appears to be self-paced with step by step guided video instruction and built-in-questions with instant feedback. During the sessions there are semi-guided "Try It!" investigation activities to let students make additional exploration and experimentation.

3.5 EV3 by ROBOESL[8]

It contains a series of curricula produced by the ERASMUS+ project ROBOESL. The material is inspired by the project-based and constructivism learning principles and proposes pedagogical approaches for robotics-based learning. The material is based on video - text and also provides validation tools to estimate the impact of the curriculum on the participants. The material is multilingual and can be used by teachers and students.

3.6 VEX IQ by VEX Robotics[9,10,11]

The material is for middle and elementary school students. There are 12 units that can be used in sequence or stand-alone lessons. The curriculum is in pdfs and contains also videos. The units contain separate content information, building instructions, rubrics, written exercises, optional activities, and teacher supplements and guides. Some of the above are free for downloading while other parts are only printed. For the development of some resources, VEX has partnered with Robomatter and PLTW, which are educational curriculum providers.

---

5 https://www.cmu.edu/roboticsacademy/.

6 https://www.cmu.edu/roboticsacademy/roboticscurriculum/Lego%20Curriculum/index.html.

7 https://www.cmu.edu/roboticsacademy/PDFs/Curriculum/Intro-to-EV3/EV3-teachers-guideWEB.pdf.

8 http://roboesl.eu/?page_id=591.

9 https://www.vexrobotics.com/vexiq/education/educational-tools.

10 https://content.vexrobotics.com/vexiq/curriculum/228-3319-VEX-IQ-Robotics-Education-Guide-20160511.pdf.

11 https://content.vexrobotics.com/vexiq/pdf/228-3428-750-Clawbot-IQ-Build-Instructions-Rev10-20150901.pdf.

3.7 EV3 by Lego[12,13,14,15]

The curriculum is developed by Lego education it is contained mostly in pdf files. The content is multilingual and follows a similar philosophy to the WeDo2 curriculum described above. It has learning goals, optional extra material list, indicative time schedules, lesson plan, student Worksheet, assessment tools along with building and programming tutorials.

3.8 Multiple Systems by Parallax[16,17,18,19,20]

The material covers a minimum of two age groups and facilitates a wide range of programming languages and robotic platforms (ActivityBot, Cyber:bot, Boe-Bot, Scribbler 3, FLiP Try-It Kit, Badge WX, Shield-Bot, ELEV-8). The content is separated by platform and programming language and follows an approximately 45–60 min lesson-based structure. The tutorials for the robotic kits are usually divided in three sections (prerequisites, main lesson, projects). The robotic platforms strongly promote the DIY movement offering additional accessories. The instructions and step by step guides are based mostly on webpages and video animations. For the registered educators it is offered a series of assessment material, extra guides, and code examples.

3.9 Edison by Robotics WPS, Microbric[21,22]

The curriculum is aligned with the technologies learning requirements of the Australian Curriculum (v8.3). The content is free downloadable pdfs and released under Creative Commons licenses thus anybody may, use, share, translate, or use it as a base to develop other lessons. The material is separated for teachers and students. It contains teacher's guides and lesson plans while for the students there are tutorials, activities, extension projects, and worksheets. The content is also separated by the programming language used. Most of the offered lessons might be completed within 90 min depending on the student's age and experience.

---

12 https://education.lego.com/en-us/downloads/mindstorms-ev3/curriculum.

13 https://le-www-live-s.legocdn.com/downloads/LME-EV3/LME-EV3_MAKER_1.0_en-GB.pdf.

14 https://education.lego.com/en-au/lessons/maker-middleschool.

15 https://education.lego.com/en-au/lessonsfilter?Products=LEGO%C2%AE+MINDSTORMS+Education+EV3+Core+Set&rows=9.

16 https://www.parallax.com/education/teach/learn/educator-resources.

17 http://learn.parallax.com/.

18 http://blockly.parallax.com/blockly/.

19 http://learn.parallax.com/tutorials?field_language_tid=All&tid=All.

20 http://learn.parallax.com/tutorials/series/activitybot-blocklyprop-tutorial-series.

21 https://meetedison.com/robotics-lesson-plans/.

22 https://meetedison.com/robotics-lesson-plans/10-robotics-lesson-plans/.

3.10 EV3 by Washington State Library[23]

It covers programming basics for the Lego EV3. It contains pdf and video files (provided in a CD). The content is broken into modules and is accompanied by 22 videos ranging from 1 min to 18. The exercise videos have no narration though some of them have occasional tips.

3.11 Arduino by STEAM Studio[24,25]

This curriculum is an accumulation of heterogeneous material from many different sources. It is mainly videos, pdfs, and websites created by individuals or companies and are presented in a single form. This approach does not require the creation of new material by the creator of the curriculum but the discovery and utilization of already available material from different sources and creators.

3.12 Multiple Systems by Innovators and Hobbyists[26,27,28,29,30,31,32]

The materials are published in "Instructables", a platform which has a community of innovators, hobbyists, and individuals with different skills who share what they are making. The material is usually webpages, with videos and external links for additional information. Usually, the material is not separated for teachers and students and does not contain material like extensive assessments, tests, timetables, etc. The "Instructable" community strongly promote open technologies (Arduino, raspberry pi, etc.) and DIY movement with step by step guides using any kind of materials. As expected, there is great differentiation between the quality of the material presented. This approach ensures new educational robotics projects on regular bases as long as the creators of the material are too many.

4 Discussion

Based on our review, it is clear that there is a great variety of ER curricula. Some of them are free while others are paid. The curricula are in many formats, like, webpages, videos, and printed materials that come with the ER products. The creators are usually the companies that develop the robotic systems (e.g., Lego, Makeblock), private or public

[23] https://www.sos.wa.gov/_assets/library/libraries/projects/youthservices/legomindstormsev3programmingbasics.pdf.

[24] http://steamcurriculum.weebly.com/arduino-based-robotics.html.

[25] http://steamcurriculum.weebly.com/arduino-microcontrollers.html.

[26] https://www.instructables.com/id/How-to-Build-the-ProtoBot-a-100-Open-Source-Super-/.

[27] https://www.instructables.com/class/Robots-Class/.

[28] https://www.instructables.com/id/Simple-Bots-Wobbler/.

[29] https://www.instructables.com/class/Arduino-Class/.

[30] https://www.instructables.com/id/Line-following-Robot-with-Arduino/#discuss.

[31] https://www.instructables.com/id/St-Patricks-Day-Pinch-Detector-With-Circuit-Playgr/.

[32] https://www.instructables.com/id/Robot-Maze-Solver/.

educational organizations, research projects, and individual educators. The curricula usually include a) user instructions for the development of simple robots or mechanisms b) user instructions for basic programming c) connection of the session with real-world problems while they rarely include a) course objectives b) separate content for teachers and students c) possible extensions and task variants d) evaluation tools.

The diversity of the curricula materials along with the lack of a general model that would support the creation of educational robotics curricula has led to a chaotic landscape where curricula rarely share common elements and principles. It is noteworthy that even among the curricula made by the same developer/company there are different deficiencies. The development of the curricula does not seem to follow a generic and comprehensive model that could lead curriculum makers to develop complete and more structured curricula in educational robotics.

Regarding the educational robotics systems and technologies, we have found that the use of open technologies (hardware/software) is making a real breakthrough in the ER curricula as teachers can adopt a low-cost, one-to-one approach for their students so each student can have or build his/her own robot to play with [34, 35]. However, the spread of open technologies is not always accompanied by the corresponding development of the necessary curricula, which makes even difficult for learners to follow user instructions to make their robots or mechanisms [36]. For open-source educational robotic constructions it is quite hard to find well-organised user instructions and corresponding curricula.

At the same time, regardless of the technologies used (open source or not), icebreaker activities are not integrated with the curricula thus teachers must develop their own icebreaking activities to facilitate their classes. Moreover, it is extremely difficult to find collaborative scripts for each session in order to help teachers to separate students into effective groups and thus accomplish the best learning outcomes. Similarly, curricula are lacking sections with detailed information on how teachers may guide students to the difficult points of a session (learning guidance or scaffolding) in order to enable learners to learn faster and more effectively without losing their interest.

Most of the robotic kits in the market are based on closed software/hardware and usually come with only user instructions to develop pre-defined robots. Although this might be beneficial as a starting point for novice users, it proposes a restricted step-by-step guided approach for learners, which often results in trivial knowledge acquisition and poor learning that does not foster creativity, inventiveness, and curiosity to be developed [37]. In contrast with this approach, the educational community proposes lately a change in educational methodologies and curricula that embraces the approach of the maker movement [38–40]. The maker movement appears to provide broad access to learning opportunities in formal and informal settings, for everyone, emphasizing mostly on the relationship between learning and making through exploration [37, 40]. The idea behind the adoption of maker and Do-It-Yourself (DIY) culture has its origins in the constructivism theory that proposes the generation of knowledge from the interaction between ideation and experience [41] arguing that learning is more effective when students have to deal with meaningful real world objects [42]. The adoption of DIY and maker culture in educational robotics suggests a paradigm shift and implies a radical change in robotics curricula. Contrary to the conventional educational robotic practices, the new paradigm encourages students to develop their own robots and robotic mechanisms using digital fabrication, arts & crafts, open-source and low-cost tools instead of using prefabricated and ready-made robots. Although, the incorporation of the maker movement sounds very attractive and has deep theoretical roots in Papert's constructionism [42] is barely identified in the existing STEAM and robotics curricula in the European schools [37].

5 Conclusion

In this article we have made a review of some typical educational robotics curricula covering many well-known technologies. We have emphasized the importance and educational value of features usually missing, such as icebreakers, collaboration scripts, learning guidance, and multilingual support which are rarely found in the ER curricula. At the same time, we have highlighted the need for a generic curriculum model that might serve as a guide for ER curriculum developers. Finally, we have argued about the need for a paradigm shift in ER curricula that will actively incorporate the maker and Do-It-Yourself (DIY) culture.

In the INBOTS project (inbots.eu) we work to this direction designing exemplary curricula for three age groups (pre-school, primary, secondary school). The curricula will become publicly available soon to promote the dialogue in the ER community and will hopefully inspire curriculum developers and teachers to contribute their own curricula or suggestions for the qualities of the robotics curricula needed for the 21st century generation.

Acknowledgements. This work was conducted by the European Lab for Educational Technology-EDUMOTIVA in the framework of the INBOTS CSA project and has received funding from the European Union's Horizon 2020 under grant agreement No. 780073 (INBOTS).

References

1. Papert, S.: Mindstorms: Children, Computers, and Powerful Ideas. Basic Books Inc., New York (1980)
2. Atmatzidou, S., Demetriadis, S.N.: Evaluating the role of collaboration scripts as group guiding tools in activities of educational robotics: conclusions from three case studies. In: 2012 IEEE 12th International Conference on Advanced Learning Technologies, pp. 298–302. IEEE (2012)
3. Blanchard, S., Freiman, V., Lirrete-Pitre, N.: Strategies used by elementary schoolchildren solving robotics-based complex tasks: innovative potential of technology. Procedia - Soc. Behav. Sci. **2**, 2851–2857 (2010). https://doi.org/10.1016/j.sbspro.2010.03.427
4. Castledine, A.-R., Chalmers, C.: LEGO robotics: an authentic problem solving tool? Des. Technol. Educ. **16**, 19–27 (2011)
5. Hussain, S., Lindh, J., Shukur, G.: The effect of LEGO training on pupils' school performance in mathematics, problem solving ability and attitude: Swedish data. In: Educational Technology and Society, pp. 182–194 (2006)

6. Sapounidis, T., Alimisis, D.: Educational robotics for STEM: a review of technologies and some educational considerations. In: Leite, L., Oldham, E., Afonso, A., Floriano, V., Dourado, L., Martinho, M.H. (eds.) Science and Mathematics Education for 21st Century Citizens: Challenges and Ways Forward, pp. 167–190. Nova Science Publishers, Hauppauge, NY, USA (2020)
7. Sapounidis, T., Stamelos, I., Demetriadis, S.: Tangible user interfaces for programming and education: a new field for innovation and entrepreneurship. In: Advances in Digital Education and Lifelong Learning, pp. 271–295 (2016)
8. Tselegkaridis, S., Sapounidis, T.: Simulators in educational robotics: a review. Educ. Sci. **11** (2021). https://doi.org/10.3390/educsci11010011
9. Sapounidis, T., Demetriadis, S.: Educational robots driven by tangible programming languages: a review on the field. In: Advances in Intelligent Systems and Computing, pp. 205–214 (2017)
10. Chlup, D.T., Collins, T.E.: Breaking the ice: using ice-breakers and re-energizers with adult learners. Adult Learn. **21**, 34–39 (2010). https://doi.org/10.1177/104515951002100305
11. Velandia, R.: The role of warming up activities in adolescent students' involvement during the English class. Profile Issues Teach. Prof. Dev. 9–26 (2008)
12. Henslee, A.M., Burgess, D.R., Buskist, W.: Faculty forum. Teach. Psychol. **33**, 189–207 (2006). https://doi.org/10.1207/s15328023top3303_7
13. Seçer, Ş.Y.E., Şahin, M., Alcı, B.: Investigating the effect of audio visual materials as warm-up activity in aviation english courses on students' motivation and participation at high school level. In: Procedia - Social and Behavioral Sciences, pp. 120–128 (2015)
14. Juang, Y.R.: Learning by blogging: warm-up and review lessons to facilitate knowledge building in classrooms. In: Proceedings - ICCE 2008: 16th International Conference on Computers in Education, pp. 279–283 (2008)
15. Fischer, F., Kollar, I., Haake, J., Mandl, H.: Scripting Computer-Supported Collaborative Learning. Springer, US, Boston, MA (2007)
16. Kollar, I., Fischer, F., Hesse, F.W.: Collaboration scripts – a conceptual analysis. Educ. Psychol. Rev. **18**, 159–185 (2006). https://doi.org/10.1007/s10648-006-9007-2
17. Rummel, N., Spada, H.: Can people learn computer-mediated collaboration by following a script? In: Scripting Computer-Supported Collaborative Learning, pp. 39–55 (2007)
18. Atmatzidou, S., Demetriadis, S.: Advancing students' computational thinking skills through educational robotics: a study on age and gender relevant differences. Rob. Auton. Syst. **75**, 661–670 (2016). https://doi.org/10.1016/j.robot.2015.10.008
19. Sapounidis, T., Demetriadis, S., Papadopoulos, P.M., Stamovlasis, D.: Tangible and graphical programming with experienced children: a mixed methods analysis. Int. J. Child-Comput. Interact. (2019). https://doi.org/10.1016/j.ijcci.2018.12.001
20. Schmidt, H.G., Loyens, S.M.M., Van Gog, T., Paas, F.: Problem-based learning is compatible with human cognitive architecture: Commentary on Kirschner, Sweller, and Clark (2006) (2007)
21. Ge, X., Land, S.M.: Scaffolding students' problem-solving processes in an Ill-structured task using question prompts and peer interactions. Educ. Technol. Res. Dev. **51**, 21–38 (2003). https://doi.org/10.1007/BF02504515
22. Forsyth, D.J.: Engaging readers and writers with inquiry: promoting deep understandings in language arts and the content areas with guiding questions. J. Adolesc. Adult Lit. **52**, 361–363 (2008)
23. Kirschner, P.A., Sweller, J., Clark, R.E.: Why minimal guidance during instruction does not work: an analysis of the failure of constructivist, discovery, problem-based, experiential, and inquiry-based teaching. Educ. Psychol. **41**, 75–86 (2006). https://doi.org/10.1207/s15326985ep4102_1

24. Clark, R., Kirschner, P., Sweller, J.: Putting students on the path to learning: the case for fully guided instruction. Am. Educ. **36**, 6–11 (2012)
25. Anewalt, K.: Experiences teaching writing in a computer science course for the first time. J. Comput. Sci. Coll. **18**, 346–355 (2002)
26. Kalyuga, S., Sweller, J.: Measuring knowledge to optimize cognitive load factors during instruction. J. Educ. Psychol. **96**, 558–568 (2004). https://doi.org/10.1037/0022-0663.96.3.558
27. Kalyuga, S.: Expertise reversal effect and its implications for learner-tailored instruction. Educ. Psychol. Rev. **19**, 509–539 (2007). https://doi.org/10.1007/s10648-007-9054-3
28. Wecker, C., Fischer, F.: Fading scripts in computer-supported learning: the role of distributed monitoring. In: Proceedings of the 8th International Conference on Computer Supported Collaborative Learning, pp. 764–772 (2006)
29. Wecker, C., Fischer, F.: From guided to self-regulated performance of domain-general skills: the role of peer monitoring during the fading of instructional scripts. Learn. Instr. **21**, 746–756 (2011). https://doi.org/10.1016/j.learninstruc.2011.05.001
30. Worsley, M., Blikstein, P.: Children are not hackers: building a culture of powerful ideas, deep learning, and equity in the Maker Movement. In: Makeology, pp. 78–94 (2016)
31. Herzog-Punzenberger, B., Le Pichon Vorstman, E., Siarova, H.: Multilingual Education in the Light of Diversity: Lessons Learned. Publications Office of the European Union, Luxembourg (2017)
32. Benson, C.: Towards adopting a multilingual habitus in educational development. In: Language Issues in Comparative Education: Inclusive Teaching and Learning in Non-Dominant Languages and Cultures (2013)
33. Olshtain, E., Nissim-Amitai, F.: Curriculum decision-making in a multilingual context. Int. J. Multiling. **1**, 53–64 (2004). https://doi.org/10.1080/14790710408668178
34. Chou, P.-N.: Skill development and knowledge acquisition cultivated by maker education: evidence from arduino-based educational robotics. Eurasia J. Math. Sci. Technol. Educ. **14** (2018). https://doi.org/10.29333/ejmste/93483
35. Saleiro, M., Carmo, B., Rodrigues, J.M.F., Du Buf, J.M.H.: A low-cost classroom-oriented educational robotics system. In: Lecture Notes in Computer Science (including subseries Lecture Notes in Artificial Intelligence and Lecture Notes in Bioinformatics), pp. 74–83 (2013)
36. Moran, R., Teragni, M., Zabala, G.: A concurrent programming language for arduino and educational robotics. In: XXIII Congreso Argentino de Ciencias de la Computación (La Plata, 2017) (2017)
37. Alimisis, D.: Emerging Pedagogies in Robotics Education: Towards a Paradigm Shift. Presented at the (2020)
38. Alimisis, D.: Educational robotics: open questions and new challenges. Themes Sci. Technol. Educ. **6**, 63–71 (2013)
39. Schön, S., Ebner, M.: The maker movement. implications of new digital gadgets, fabrication tools and spaces for creative learning and teaching. Elearning Pap. **39**, 1–12 (2014)
40. Blikstein, P.: Digital fabrication and ‘making’ in education the democratization of invention. In: FabLab. transcript Verlag, Bielefeld (2014)
41. Piaget, J.: To understand is to invent: the future of education. In: International Commission on the Development of Education (1973)
42. Papert, S., Harel, I.: Situating constructionism. Constructionism. 1–12 (1991)

Dance and Robots: Designing a Robotics-Enhanced Project for Dance-Based STEAM Education Using ENGINO

Sofia Almpani[1,2(✉)] and Dimitris Almisis[1]

[1] EDUMOTIVA, 23100 Sparta, Greece
info@edumotiva.eu

[2] School of Electrical and Computing Engineering, National Technical University of Athens, Athens, Greece
salmpani@mail.ntua.gr

Abstract. The emerging domain of robotics provides a creative tool for engaging young children with the engineering and artistic components of STEAM. Educational activities based on dance can captivate student interest in arts while exploring foundational technological ideas. The paper describes vignettes of dance-based activities utilizing the ENGINO E30 kit. The objective of this work is to explore ways to use a robotic kit to create a science-art integrative STEAM project, named DancENG. The DancENG project has been designed by creatively integrating dance, music, engineering, and math concepts, aiming to cultivate aesthetic sensitivity in parallel with problem-solving and engineering-thinking skills of elementary school students. It has been implemented to 20 5th graders of a Greek elementary school to indicate the practical use of the framework.

Keywords: Education · STEAM · Mathematics · Art · Dance · Educational robots · ENGINO

1 Introduction

Several studies show how STEM education, a multidisciplinary integration of Science, Technology, Engineering and Mathematics, enhances the development of convergent abilities in students curriculum [1, 2]. Art in its various forms (painting, music, theater, dance, etc.) can provide different innovative ways to interact with STEM and expand it into STEAM (A for Art), increasing the efficiency of learning and motivation as depicted in recent research [3–5]. However, art-based STEAM program has not been diversely established because Arts has only played a secondary or marginal role and, moreover, there are only a few cases concerning the utilization of emerging technologies as robots in science-art integrative education [6, 7]. The robotics-enhanced project for art-based STEAM education suggested in this study is expected to provide educational tools to enhance creativity and ingenuity through the natural integration of arts and robotics.

Among these art forms, dance, as an embodied cognitive medium [8], constitutes an expressive approach for creating educational activities including STEAM disciplines

M. Malvezzi et al. (Eds.): EDUROBOTICS 2021, SCI 982, pp. 139–151, 2021.
https://doi.org/10.1007/978-3-030-77022-8_13

[9–12]. Dance can be depicted as a social event that expresses artistic and communicative elements through movement, helping in releasing experimental students' inspiration and transferring creative thinking and behavioral outcomes into everyday environments [3]. Thus, dance-based educational activities emerge as appropriate inventive method in the comprehension of concepts from math, physics, engineering, etc. [3, 11]. There is no doubt that art can play an important role when exploring technology and science, but how teachers combine these domains effectively can be a challenging task. Especially during the elementary school years when educators are already dealing with a large load of teaching curriculum [13].

One of the aims of this paper is to explore how robotic kits (like ENGINO) can facilitate STEAM education using dance activities in a natural way that can easily integrate with traditional subjects [13]. Additionally, STEAM education has a place in the proposal of solutions to overcome the well-studied and identified gender concerns, such as absence of female role models, relevancy of curricula, socio-cultural beliefs and expectations [14], aiming to encourage and retain females in the domains of science and technology. The same applies for other often-excluded groups of students, such as immigrants, lower social-economical groups, disabled.

To this end, our work attempts to aid towards establishing a more inclusive formal and informal STEAM education by suggesting **DancENG** (abbreviation of Dancing ENGINO) approach that is meaningful not only for technological-oriented students (mostly boys) but also for students with various levels of technical ability and interest such as girls, immigrants, students with disabilities, etc. DancENG is a proposal for a STEAM environment which includes dance performances where educators, students, and technologies work together to implement artistic tasks to learn STEAM concepts [3] through hands on, collaborative projects with educational robots [15].

2 DancENG Educational Framework

Composing dance choreography is very similar with problem solving, including notions such as ideas, experimentation, patterns, holistically organization of the structure and form [16]. A meaningful assumption of the DancENG approach is that the incorporation of specific supporting technologies enhances knowledge construction and learning procedures within a STEAM curriculum. In this context, DancENG attempts to create the methodological framework for the realization of dance-based activities supported by educational robotic kits that use dance sequences as embodied STEAM notions to cultivate the creation of parallel abstract notions. The body and its movements have a pivotal role in the DancENG project, both as the artistic discipline of dance and as the physical embodiment of the robotic technology.

In this study, ENGINO E30[1] robotic kit was used as a representative example of educational robotic kits for primary and secondary school students [17]. ENGINO models are built by assembling small blocks or bricks together, with the aim of assisting students to create technological models easily and creatively so that they can explore and learn in a playful way [17]. Studies [18] have validated that the ENGINO products create the

[1] https://www.engino.com/.

suitable circumstances for developments of creativity and novelty, indicating that the ENGINO kit is well-suited for exploring a variety of STEAM concepts. Even though there are many remarkable educational robotic kits that can be used in this project, they are mostly divided into two categories. They are either constructional kits to build your own models, but they require computer programming, such as LEGO WeDo, VEX, Popy Ergo Jr, etc. [19, 20], or they provide tangible programming but they are ready-made robots, such as Probot, Bee-Bot, etc. [20, 21]. ENGINO E30 kit combines both since it is a constructional kit offering tangible programming. Another example that we can say that it also combines the two categories is Cubelets [20], but cubes restrict the variety of models and its tangible programming doesn't provide the required flexibility. However, DancENG project can be also implemented with other educational robotic kits after some necessary modifications according to the technology used.

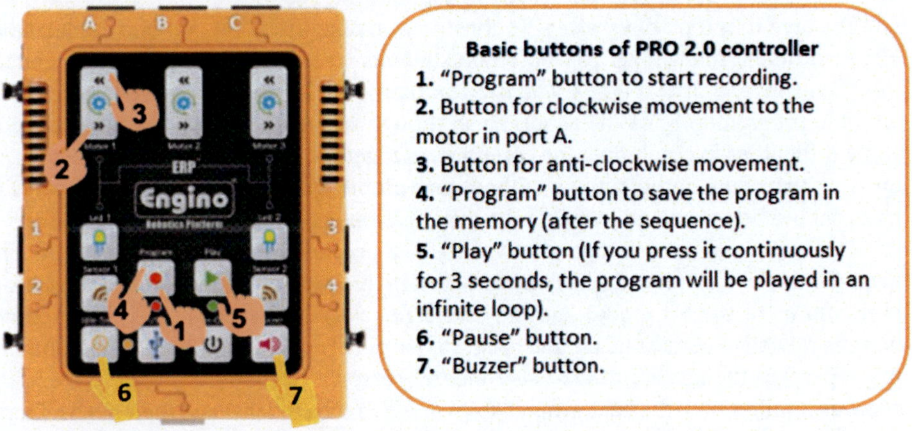

Fig. 1. Basic buttons of ENGINO PRO 2.0 controller for manual control and record.

These are some of the important advantages of ENGINO robotic kit that makes it suitable for the proposed project. First of all, ENGINO's technology has the ability with very few components to allow the connection in all directions of the three-dimensional space facilitating the students to build very easily and quickly their own models as well as to modify them within a few minutes. This allows a great degree of possible variations. This degree of variations enhances students' creativity and engineering design thinking. Additionally, the "PRO 2.0 controller" (see Fig. 1) provides tangible and unplugged programming, which is preferable for the age of elementary school students and it is more appropriate for the dance-based approach of this project, since the no-screen programming permits the unhindered movement all over the space[2]. Furthermore, the "buzzer" and "pause" buttons provide a perfect tool for the teaching of notes length, pauses, and measures (Activity 2) and the "play" button allows playing a sequence in an infinite loop which is well-suited for the repetition of dance sequences.

[2] The "PRO 2.0 controller" offers also other interconnected ways of programming, code, flow diagram, etc. for more advanced programming based on the children's age and experience.

In the implementation of the proposed project participated 20 children aged 10 years old that were students in the fifth-grade class of a public elementary school in Athens, Greece. The children were grouped in teams of 3–4 pupils. A few children had some previous experience in dance (mostly girls) or in robotics (mostly boys), but generally the class had little experience in the STEAM objectives that this project introduces. The pilot project lasted two weeks (June 2020) and its duration was approximately 10 h, two hours per activity 1–3 and four hours for activity 4 (of course the time duration can vary and it depends on pupils' level and the complexity of dance combinations. Activities can repeat with more complex combinations).

The proposed activities are team-based and the teacher has mostly a supportive role. This means that in the beginning the teacher describes the activity, the concepts included and what pupils should do, and then each team works independently. The teacher moves from group to group and only intervenes when a team needs help. At the end, each group presents to the other groups the results of their work.

The activities took place after the construction of an ENGINO robotic model, which can be different in each activity based on students' preferences. In the implementation the "Bumber car" (see Fig. 3) was used utilizing the "PRO 2.0 controller" for direct tangible programming of the robot. Even though in the beginning the students found some difficulties in combining the components together, after a few minutes they started creating their own combinations without any problem and they explored the ways the parts should be connected to create the desirable outcome.

The mathematical concepts and vocabulary practiced in the sample activities had been previously introduced (the concepts used in this project are included in the national curriculum[3] of the 5th grade) and they were reviewed, enhanced, and connected with notions related to dance and engineering. In our pilot implementation, the first author of the paper was the teacher of the class and the person[4] that carried out the project, thus she was familiar with the background knowledge of the students and their needs. During this project they had the chance to revise the acquired knowledge and enhanced it with the new concepts creatively.

In the following section, the illustrative activities for the proposed STEAM framework are presented.

3 The Proposed STEAM Activities of the DancENG Approach

DancENG proposal aims to link concepts from dance with maths and engineering within a unique educational framework. Some basic elements of dance [16, 22, 23] used as a base in this proposal are:

- **Space**: *where* is movement performed, related to directions, floor pattern, etc.
- **Time**: *when* is movement performed, related to rhythm, duration, phrases, etc.
- **Energy**: *how* is movement performed, related to force and flow.

[3] Greek Curriculum – Primary School – Mathematics: https://repository.edulll.gr/1926.

[4] The teacher had previous experience with STEAM-related teaching (dance and math), but if it is possible, a multi-disciplinary team of educators is also proposed.

- **Form**: how is dance *structured*, related to recurring theme, abstract geometrical form, repetition, and other choreographic devices.

In the DancENG activities below, the description and the didactic implication are described, as well as the related topic and the goals [24] of each discipline in Table 2.

3.1 Activity 1: Move in the Space

Description/Implication: In this activity, pupils explore the classroom's space, directions, and floor patterns [16] by creating lines and shapes through movement, first by themselves (following the context of embodiment [8]) and then by programming the robot to do the same movement. Children mention the different shapes they are familiar with and practice moving in *straight lines, squares, circles, butterfly-shape*, etc. in the classroom (see Fig. 2), which is a common practice in dance lessons for understanding of space. Then, they examine which of these shapes ENGINO robot can or cannot do with its controller's programming language by testing and correcting the engineering design process to make the robot move in the different shapes and even in curves and circles, either all over the class space or by drawing them in a paper after adjusting a pencil on ENGINO robot.

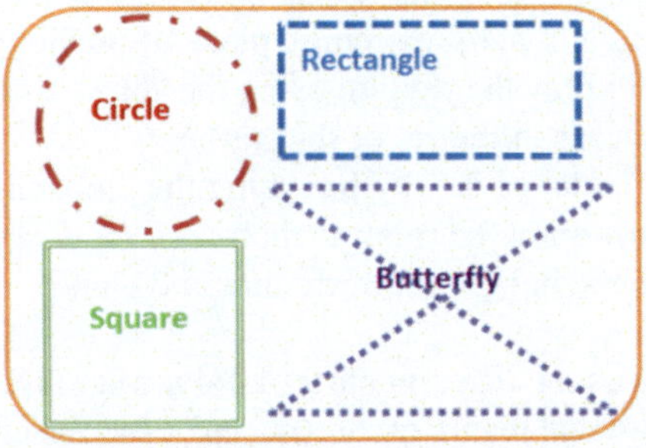

Fig. 2. Shapes from "Move in the space" activity.

During the pilots, students managed to demonstrate all the proposed shapes after several trial and error attempts and they had fun by competing each other team on who will be the first to succeed. One of the important things that they had to explore is which buttons and for how long should they press in order for the robot to turn (e.g. it requires a 90° turn for the square and a 45° turn for the butterfly) The robots can provide instant feedback—they either do the correct shape, or they do not. Making mistakes and troubleshooting are fundamental steps for developing problem-solving skills thus they provide a unique learning opportunity. The dance practice also provided an important help to understand how the robot should move analogously with their own body movement (in terms of embodied programming).

Note that in all activities, once the sequence of the program has been inserted manually by pressing the onboard offline keyboard on the PRO 2.0 controller, it is saved on the robot instantly. No connection with a computer or tablet is necessary for programming with ENGINO (however, ENGINO provides the ability to also program the robot with a computer/tablet/smart phone, using a simulator or a flow diagram or code, enabling this project to be suitable for older students as well).

3.2 Activity 2: Feel the Rhythm

Music is an essential part of dance lessons, which require understanding of concepts such as **beats, bars (or measures), notes length, and rhythm** [22]. Music is divided into beats and bars. Beats are similar to single dance steps, while taking a few beats together creates a bar, just like a few steps give you a pattern. Bars combine to give phrases, such as a few dance patterns build a small choreography.

Rhythm work in dance most of the times use a **4/4 time signature** and the steps are usually counted up to 8, i.e. *two bars counting by 1,2,3,4 - 5,6,7,8.* Music is read in a similar way as math symbols. The lengths of notes are indicated by mathematical **fractions**. Each note has a different shape to indicate its beat time. There are whole notes, half notes, quarter notes, and eighth notes. Teaching notes length is the same as teaching math fractions, thus they can be combined (see Table 1).

After teaching the notes length, the teacher gives to the pupils a set of rhythm, (*e.g., a whole note, two crochets and a minim*) and they can do the following variations:

1. Clap the set of rhythm to a 4/4 marching piece of music.
2. Step the set of rhythm to a 4/4 marching piece of music.
3. Program the robot to play the rhythm using the "buzzer" button (or also the "pause" button if there is pauses in the set of rhythm).
4. Program the robot to dance the rhythm using the "movement" buttons.
5. A group of pupils program the robot to their own set of rhythm and the other groups of pupils try to understand the set of rhythm and clap it.

The teacher can try using different sorts of 4/4 music, different speeds, and different styles (such as jazz, classical piano, orchestral, etc.) and the children should try not only to play the rhythm by themselves, but also program the robot so that it follows the rhythm correctly. Pupils can also propose their own song choices.

In the implementation, there were some difficulties in following the correct counting of the music (especially in songs with quick tempo), which required spending some time in the beginning of the activity only to count the music into 2 bars. Students really enjoyed this inter-disciplinary activity, which combined elements of dance, music, math and technology, and give them an insight of how different disciplines can be closely related. The variation 5 is proposed to be done after working for 2–3 times the other variations.

Table 1. Combination of fractions, note lengths and dance in the activity "Feel the rhythm".

Activity 2: "Feel the rhythm"			
	Dance counting	Math Fractions	
Semibreve A Semibreve is *whole note* (= four counts).	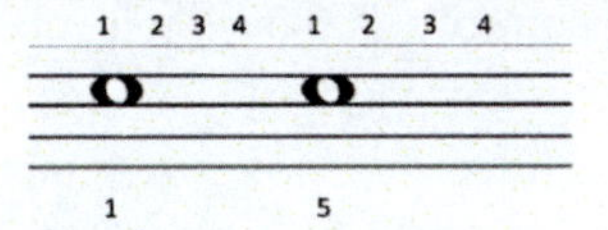	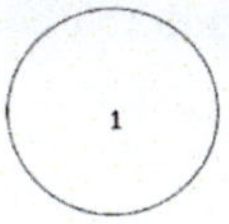	$\frac{1}{1}$
Minim A minim is a *half note* (= two counts).	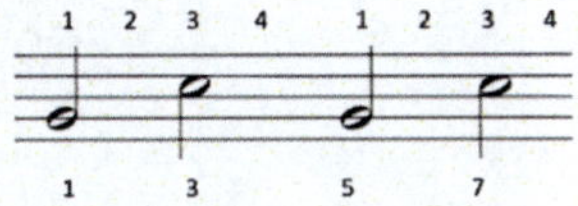	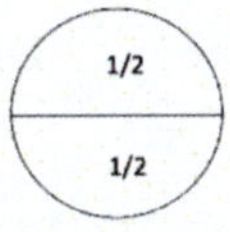	$\frac{1}{2}$
Crochet A crochet is a *quarter note* (= one count).	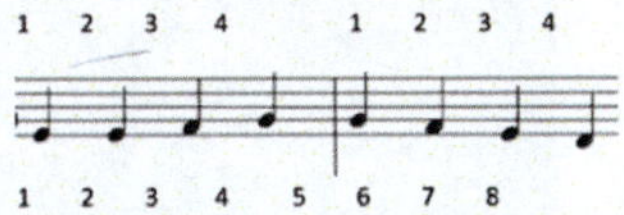	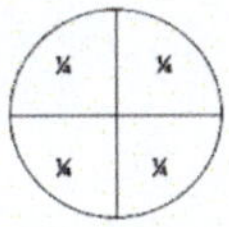	$\frac{1}{4}$
Quaver A quaver is an eight note (= half note).		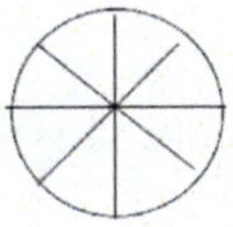	$\frac{1}{8}$

3.3 Activity 3: Synchronize Rhythm with Dance

In this activity, the children synchronize a small dance sequence on the set of rhythm that they learned in Activity 2 and then they "teach" the same sequence to the robot. Based on children's dance level the dance sequence can be created with the following variations. In all levels, **first the pupils learn the sequence to dance it by themselves and then they teach the same sequence to their robots**, aiming in both cases to synchronize the dance sequence with the rhythm of music.

- *Level 1*: The teacher provides to the pupils a pre-arranged dance sequence.
- *Level 2*: The pupils develop the dance by putting a floor pattern (from Activity 1) on simple steps of the rhythm. Pupils must count the note values as they do each step. The teacher makes sure they can repeat it. Then they add arm movements on the dance steps or even add other steps (such as turns or leaps). This level introduces the student to a more creative approach. The movement of the robot should be kept simple (it cannot jumb or use hands!).
- *Level 3*: The children make their own combination. They think of 3–4 dance steps (including arms, directions, counting, etc.), write them down together with their note lengths, and dance them out.

Dancing the sequence helps the children to separate a complex task into smaller, executable steps, which is similar with the sequential nature of programming [25]. Programming involves specifying each movement independently in the right order. When the children have captured the dance sequence, they try to "teach" the dance sequence to the robot (mostly regarding to floor patterns, note lengths, and simple movements). For example, in the set rhythm used as example in Activity 2, they can program the robot to do the following sequence. "Move forward for four counts, turn left for two counts, move backward one count, and turn right for one count." Teaching the robots to dance, teaches children how to control and move their robot (Fig. 3).

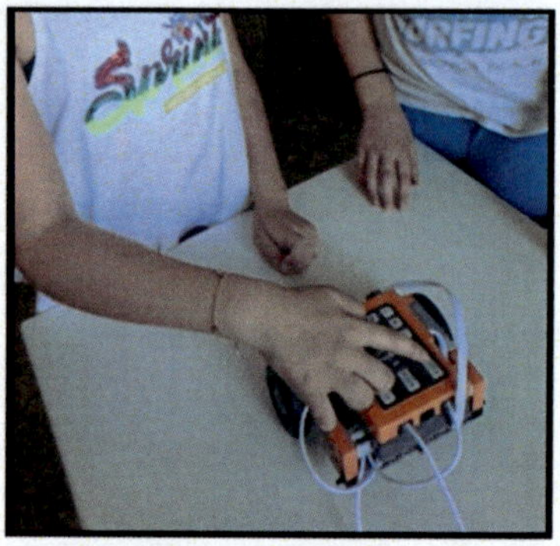

Fig. 3. Pupils "teaching" a dance sequence to the "Bumber car" from the pilot implementation.

Dance sequences usually repeat, so making the robot to repeat a sequence of executable steps is the next stage. With a loop, the robot can repeat the sequence. The ENGINO robot, by simply pressing the "start" button for 3 s, repeats the last sequence in an infinite loop. By doing so, it helps children explore foundational programming concepts such as sequence and loops, and engages them in robust programming and mathematical learning.

A series or sequence that repeats creates also a math pattern. Besides geometry and time, everything in dance has to do with patterns. Students memorize patterns as they act out dance sequences executed in the space. Additionally, the set of rhythm contains patterns in the form of beats, which is usually synchronized with the dancers' movement. Indicating this connection in class, can make the students see and understand mathematics as a part of everyday life [10].

In the implementation, the class made a few examples all together (practicing both students' movements and robot's movement) and they proposed what moves can be added to each note length. A main difficulty in this exercise is the "turn" movement, since it takes specific time for the robot to turn. In this case we either use a slow tempo and we count how much time it takes for the robot to make a full turn or we agree with the students that the robot will turn as much as the note length permits (e.g. half circle). Additionally, in level 2, if we want the robot to follow the floor pattern we should keep two empty bars between the repetitions of the set rhythms in order to change direction. Generally, keep the dance sequences very simple in the beginning (it can be just steps) and level up only if the pupils have mastered the previous level.

The quick and tangible way that ENGINO can be programmed to repeat loops was very helpful to show the difference between a single sequence and a repeating one. In the pilot project, children spontaneously played a game where they were trying to guess in which direction the robot will move during the loop and see if it was going to find obstacles in the class (like tables and chairs). Generally, during this project, pupils can and should be encouraged to find their own creative alteration of these activities.

3.4 Activity 4: Make Your Choreography

In the last activity, pupils became choreographers and directors as part of their STEAM exploration [13]. Children chose songs from the class's music collection and programmed their robots to complete dance sequences synchronized with the songs' tempo. After their robot was arranged in the way they desired, pupils choreographed their own movements to dance out along with the robot. Some of the elements that they should take into consideration are the synchronization of dancers, the directions in space, and the rhythm of music. The outcome of this work can then be shared with other students of the school or demonstrated to parents, since the main concept of dance is to communicate thoughts, feelings, and ideas. In our pilot, the teams demonstrated their choreographies between them as well as to the other fifth-grade of the school.

Fig. 4. Dance choreography with human and robot dancers from the pilot project.

During this activity, children explored features of stage production, as they experienced being live performers, directors of the dance show, and even the engineers who checked that the technology (i.e., the robot and program) are all set for the performance (Fig. 4) [13]. The idea of integrating dance and stage performance into robotics education provided the potential to accelerate the development of problem solving, communication and critical thinking skills in pupils through the procedures of *inspiration, planning, creation* and *performing* of a dance play in parallel with the *programming* of robots as dancers.

After the activities, the pupils were asked on their impressions from the project (the intentions of this assessment weren't on the learning goals rather on their experience). The general impression was that the pupils had the chance to cooperate, to solve together the problems that they encountered, to be creative and to have fun at the same time. To quote the words of a student (Fig. 5):

> *"I liked that we collaborated and spent our time creatively, that we assembled bricks, that we laughed at the moves made by the car we made (i.e. the robotic "Bumber car"), and that we made a chicken with my team. It was difficult for me to assemble the bricks because they need a specific way. General, I did not find anything else difficult. I would like to make a cat or a dog that dances in different directions."*

While in the current state, this implementation included a small number of pupils, future work is required that will involve more students and more explorations to test and validate the impact of the DancENG framework.

Table 2. Topic (T) and goals (G) of the activities in connection with STEAM fields.

Activities		STEAM fields		
		Dance	**Mathematics**	**Engineering**
1. Move in the space	T	Space	Geometry - Shapes	Basic engineering
	G	-Experiment and identify how the different movements define the use of space in dance.	-Recognize standard concepts as shapes. -Imaginary drawing and real drawing of geometrical shapes.	-Learn the basic programming actions and the procedure of commands and program the robot.
2. Feel the rhythm	T	Rhythm	Fractions	Basic programming - Sequencing
	G	-Learn about rhythm, beats, bars and tempo and note lengths and how they are used in dance. - Exercise different set of rhythms.	- Understand a fraction 1/b as the quantity formed by 1 part when a whole is partitioned into b equal parts.	- Make a "buzzer" and "pause" sequence. - Combine the note length and pauses (time) with the robotic movements' commands.
3. Synchronize	T	Synchronization	Time	Sequencing - Repeat Loop
	G	- Synchronize the dance movements with the set of rhythm and develop musicality.	- Understand that time is used to quantify, or measure the duration of (dance) events or the intervals (pauses) between them and even, sequence of events.	- Make the robot to repeat a sequence of executable steps. - Combine key computational concepts as identifying phases and orders within a set of rhythm.
4.Choreography	T	Choreography	Patterns	Combined program
	G	- Divide a task into a series of interconnected sub-tasks. - Create a performance, including music, dancers and robots.	- Learn progression patterns, by making alterations of the same sequence.	- Develop the ability to program group lines of code into a unit and combine different parts in a whole.

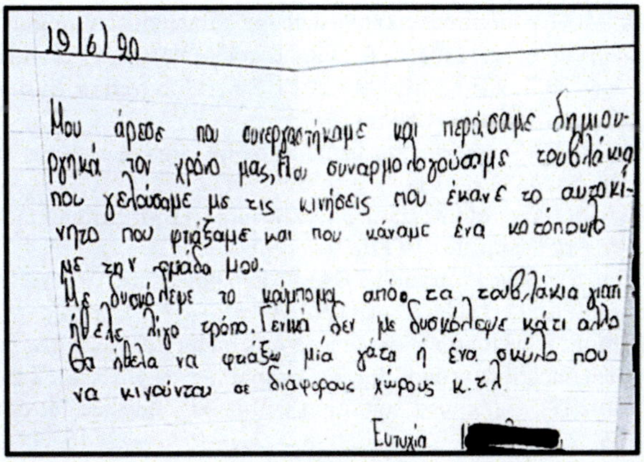
19/6/20

Μου άρεσε που συνεργαστήκαμε και περάσαμε δημιουργικά τον χρόνο μας, που συναρμολογούσαμε τουβλάκια που γελούσαμε με τις κινήσεις που έκανε το αυτοκίνητο που φτιάξαμε και που κάναμε ένα κατόρθωμα με την ομάδα μου.
Με δυσκόλεψε το κάμπωμα από τα τουβλάκια γιατί ήθελε λίγο τρόπο. Γενικά δεν με δυσκόλεψε κάτι άλλο
Θα ήθελα να φτιάξω μία γάτα ή ένα σκύλο που να κινούνται σε διάφορους χώρους κ.τ.λ

Ευτυχία [redacted]

Fig. 5. The experience of a student from DancENG project (in Greek).

4 Conclusions

In this paper, an original proposal for a STEAM educational framework that combines dance and robots, called DancENG, has been presented. The aim was to show that the introduction of dance (and art in general) in teaching procedures of STEAM can enhance embodied learning, increasing at the same time students' creativity, motivation, and engagement. We have also proposed that the use of robotic kits (such as ENGINO E30 kit) as a technological medium for the engineering part in the accomplishment of the activities, offers unique possibilities for artistic, mathematical, and engineering learning. Assembling and programming a robot encourages children to use their creativity and engineering-thinking skills in a playful, engaging learning environment. While this project was tested with a small group of pupils, future work is needed that will involve the further implementation and experimentation with more students and classes to check and validate the impact of the suggested STEAM framework. In the INBOTS project[5], we work towards a highly accessible and sustainable framework in robotics education [26] to promote the broader use of robotics as a tool to teach and learn STEAM subjects and the DancENG approach comes to contribute to this end.

Acknowledgment. This work has received funding from the European Union's Horizon 2020 research and innovation program under grant agreement No. 780073 (project INBOTS CSA).

References

1. Cotabish, A., Dailey, D., Robinson, A., Hughes, G.: The Effects of a STEM Intervention on Elementary Students' Science Knowledge and Skills, School Science and Mathematics, vol. 113 (2013)
2. Kanematsu, H., Barry, D.M.: STEM and ICT Education in Intelligent Environments, vol. 91, pp. 3–198. Springer (2016)

5 https://inbots.eu/.

3. Città, G., et al.: Move your mind: creative dancing humanoids as support to STEAM activities. In: De Pietro, G., Gallo, L., Howlett, R., Jain, L. (eds.) Intelligent Interactive Multimedia Systems and Services 2017. KES-IIMSS-18 2018. Smart Innovation, Systems and Technologies, vol 76. pp. 190–199. Springer, Cham (2017)
4. Land, M.: Full STEAM ahead: the benefits of integrating the arts into STEM. Procedia Comput. Sci. **20**, 547–552 (2013)
5. Land, M.: The importance of integrating the arts into STEM curriculum. In: Converting STEM into STEAM Programs, pp. 11–19 (2019)
6. Kim, J.: Study on SW education and STEAM education using Arduino for technology subject. Korean J. Technol. Educ. **15**(1), 22–48 (2015)
7. Kim, J.-O., Kim, J.: Development of science-arts integrative STEAM program in elementary school using the educational robot. Int. J. Comput. Sci. Inf. Technol. Educ. **1**, 47–52 (2016)
8. Shapiro, L., Stolz, S.: Embodied cognition and its significance for education. Theory Res. Educ. **17**, 19–39 (2019)
9. Champion, D.N.: The STEAM dance makerspace: a context for integration: an investigation of learning at the intersections of STEM, art, making and embodiment. ProQuest LLC, Ph.D. Dissertation, Northwestern University (2018)
10. Mannone, M., Turchet, L.: Shall we (math and) dance? In: Mathematics and Computation in Music, 7th International Conference, MCM 2019, Madrid, Spain, 18–21 June (2019)
11. Walker, K.: Dance – a healthy gateway to STEAM practice. STEAM J. **3**, 1–3 (2018)
12. Wasilewska, K.: Mathematics in the world of dance. In: Proceedings of Bridges 2012: Mathematics, Music, Art, Architecture, Culture, pp. 453–456 (2012)
13. Sullivan, A., Strawhacker, A., Bers, M.: Dancing, drawing, and dramatic robots: integrating robotics and the arts to teach foundational STEAM concepts to young children. In: Robotics in STEM Education: Redesigning the Learning Experience, pp. 231–260 (2017)
14. Blickenstaff, J.C.: Women and science careers: leaky pipeline or gender filter? Gender Educ. **17**, 40369–18 (2005)
15. Suero Montero, C., Jormanainen, I.: Theater meets robot – toward inclusive STEAM education. In: Advances in Intelligent Systems and Computing, vol. 560, pp. 34–40 (2017). https://doi.org/10.1007/978-3-319-55553-9_3
16. The 4 Basic Elements of Dance Composition. https://mskiddsjam.weebly.com/. https://mskiddsjam.weebly.com/uploads/3/9/0/7/39078883/atc_1oi_-_elements_of_dance.pdf. Accessed 22 Sept 2020
17. Antoniou, E.N., Araújo, A., Bustamante, M.D., Gibali, A.: Physically feasible decomposition of engino toy models: a graph theoretic approach. Eur. J. Appl. Math. ArXiv (2017)
18. Danos, X., Constantinou, C., Livitzis, M., Avraam, C.: Action research to develop a scheme of work to promote creativity and designerly thinking through play. Design Technol. Educ. Int. J. **19**, 17–29 (2014)
19. Davcev, K., Koceska, N., Koceski, S.: A review of robotic kits used for education purposes. In: X International Conference of Information Technology and Development of Education ITRO (2019)
20. Komis, V., Misirli, A.: The environments of educational robotics in Early Childhood Education: towards a didactical analysis. Educ. J. Univ. Patra*s* UNESCO Chair (2016)
21. Sapounidis, T., Demetriadis, S.: Educational robots driven by tangible programming languages: a review on the field. In: Advances in Intelligent Systems and Computing, International Conference EduRobotics, vol. 560, pp. 205–214 (2016)
22. Barker, N.J.: ELEMENTS OF DANCE. Bulacan State University-Malolos Study Resources (2007) https://www.coursehero.com/file/32632124/dance-elementspdf/. Accessed 22 Sept 2020
23. Hackney, P.: Making Connections: Total Body Integration Through Bartenieff Fundamentals (2003)

24. Conklin, J.: Review of a taxonomy for learning, teaching, and assessing: a revision of bloom's taxonomy of educational objectives complete edition. Educ. Horiz. **83**(3), 154–159 (2005)
25. Francis, K., Poscente, M.: Building number sense with Lego Robots. Teach. Child. Math. **23**, 310 (2017)
26. Alimisis, D.: Emerging pedagogies in robotics education: towards a paradigm shift. In: Pons, J. (eds.) Inclusive Robotics for a Better Society. INBOTS 2018. Biosystems & Biorobotics, vol. 25. Springer, Cham (2018)

Robots Entering the Care Sector. The Case of a New Curriculum for the Education of Assistant Nurses in Sweden

Britt Östlund(✉)

Royal Institute of Technology, KTH, Hälsovägen 11C, 14157 Huddinge, Sweden
brittost@kth.se

Abstract. Robots are entering care work but the development remains slow. One key issue is how to educate healthcare professionals in the use of robots. This paper explores robots in relation to the education of assistant nurses working in care for the elderly. How can we make robots part of the curriculum and what is it that future care workers really need to know? A collaboration between the Royal Institute of Technology, KTH, and 49 teachers at the College of Care and Welfare, VO-COLLEGE, in Sweden that includes teachers, two seminars, two course development meetings and one design workshop aims to explore teachers' experiences of technology in care for the elderly work and in teaching technological matters. The results show that design workshops seem to be a more useful method to generate suggestions on how robots can be better adapted to care work, compared to seminars that brought up broader, though important, questions about care work and the role of assistant nurses and care receivers. The conclusion is that a greater awareness of technology used in care for the elderly is important to contribute to a relevant adaptation of robots and to be able to meet the challenges of an increased demand for care and care professionals. Increased responsibility for this development conveyed in education can help to broaden the view of technology, from being passive recipients of technology to being active co-creators.

Keywords: Care robots · Digitalization · Assistant nurses · Educating assistant nurses

1 Introduction

This short paper is a presentation of an ongoing development of a curriculum for assistant nurses in Sweden, more precisely an attempt to find out what the students need to know about robots when they complete their education. The case is based on a collaboration between the College of Care and Welfare in Sweden (VO-College) and the Royal Institute of Technology (KTH), in Stockholm Sweden 2017–2019, include seminars for teachers at VO-College provided by KTH, and discussions with teachers at VO-College on how they convey knowledge on IT in care and welfare.

Robots create new opportunities for how we can provide care. Being part of the broader framework of digitalization, robots are perhaps one of the most thought-provoking applications and expectations are high of what they can bring to virtually

M. Malvezzi et al. (Eds.): EDUROBOTICS 2021, SCI 982, pp. 152–163, 2021.
https://doi.org/10.1007/978-3-030-77022-8_14

every area of society [1, 2]. Compared with other sectors in society, care is something more than nursing and offers special challenges. The adoption and technology readiness of available robot applications are high but a more thorough contextualization is needed to broaden the uptake [3]. International reports indicate that digitalization has great potential for automating working life, but the differences between sectors is large. The potential for further automation of tasks in industry is large, while the potential is significantly lower in healthcare, actually 36% compared to 60% for manufacturing or 73% for the restaurant sector [3–5]. Of course, working with people places completely different demands on flexibility and sensitivity. The question of automation here is the use of robots and how they should be organized and used, which are among the key issues for educators in healthcare. Another key question is how to integrate robots in the education of professional care workers. How to make robots part of the curriculum and what is it that future care workers really need to know? Also, can robots be part of teaching care workers, or at what stage will they experience this technology?

Assistant nurses constitute the largest occupational group in the Swedish labor market and is dominated by women, at close to 90% [20]. The need for trained assistant nurses is expected to increase by 170,000 by 2035 [20].

The majority of assistant nurses work in municipalities but they also work in hospitals and nursing homes. One difference between nurses and assistant nurses is that nurses have credentials, which assistant nurses do not have, even though they have at least 18 months of education. The tasks performed by assistant nurses consist of helping with daily hygiene and household duties, such as washing dishes, doing laundry, cleaning, supervising the delivery of prepared food, providing healthcare such as changing dressings and distributing medicine and providing social support [12].

Employing assistant nurses and care is the responsibility of local governments and municipalities in Sweden, while visions of digitalization and national strategies are the responsibility of the government [9, 10]. Evaluations 2015, 2017 and 2020 of initiatives to digitalize in the municipalities show that the developments are slower than expected [9, 11].

The use of technology is well known to care workers in Sweden and has been an important part of the emergence of the modern home help services and clinical caring activities. This development includes personal emergency response systems (PERS) implemented in the 1970s and now being digitalized; through to the 1990s IT-development with computer terminals for home shopping and documentation; to digitalization with care robots and monitoring [13]. As a result of the ongoing technological development in care for the elderly several initiatives within the framework of assistant nurse education are starting up [8]. Recently, in 2019, the government launched a new revised curriculum for assistant nurses but without a special focus on technology. The curriculum encompasses 18 months with the goal of giving the students' knowledge in health, disabilities, perspectives on ageing and lifestyle as well as the historical developments in healthcare. The new curriculum also includes the ability to exercise social skills and communicate needs of the care receivers and to take an active part in planning care work, working safely, and suggesting improvements in the work environment.

VO-College is a platform for collaboration between employers, trades unions and educators in the care sector with the aim of providing high quality education expressed in their own certification of the participating trainers. Teachers at VO-College realized that the need for learning about digitalization of care work was urgent. They were especially interested to discuss robots as part of digital developments. One reason was that while digitalization is increasing in the care sector, the technology aspects are still virtually non-existent, with the exception of a couple of courses with a special focus on welfare technology [7]. In particular, the skepticism often mentioned in the Swedish debate, that robots in education are talked about but are rarely part of education systems or curricula [8].

VO-College initiated a contact with KTH 2017 to request training for teachers educating assistant nurses. The long-term goal was to develop a curriculum on technology or integrate technology into the current curriculum. KTH is the main technical university in Sweden and has recently adopted a new research path, which added a technology and social change perspective to medical engineering.

1.1 Objectives

The aim of this paper is to explore robots as part of teaching assistant nurses on how to use robots in care work and to reflect upon how to contribute to relevant design. The paper explores what kind of questions should be included in a future curriculum on digitalization for assistant nurses and what difference it makes to participate in the design of robots to be used in care work. The approach is thus, what has been defined as a "theme-based curriculum approach" where the use of robots in home care work is the special topic for learning [6]. For KTH, this was a generic process in which robots were assigned a wider range of uses than the scenarios imagined in laboratories.

Educating future healthcare professionals faces several challenges, in Sweden as well as on a global scale: new knowledge (i.e. epidemiology), new technology, i.e. digitalization demanding new ways of working and a better knowledge of citizens demands and needs [19]. There are also high expectations of what digitalization can achieve in care for the elderly. Ever since the first Information technology (IT) visions were launched in a global and European context, older people have been identified as those who had the most to gain from this development [21] followed by national IT strategies, commissions and research funding. In the Swedish government's strategy for Life science and health, welfare technology is highlighted as one of the spearheads of digitalization, which, through participation from the elderly, and other citizens in need of help, will lead to increased independence and better health [22]. The general argument for these initiatives has been to meet the growing needs of a growing elderly population. Against this background, the *first objective* was to examine teachers' expectations, and in doing so in comparison with policies published, with the ambition to increase the quality of care for the elderly with technology.

The *second objective* was to get an overview of existing practices and teaching materials. Even though technology is just at the start of being made part of general assistant nursing education, it is important to know how teachers are actually addressing this issue already.

The *third objective* is aimed at robots in care practice. Since internship and practicing care for the elderly is a big part of assistant nursing education, teachers as supervisors study students' experiences and any use of robots or digitalization taking place. In combination with the long use of technology in care for the elderly, it can be assumed that these experiences will also affect the adoption of future technologies [36].

The *fourth objective* was to examine the design and usefulness of robots in care for the elderly practice. Using a design workshop method this was an exercise to actually be pro-active and experience the role of not being a passive recipient of new technology but instead to explore teachers' own ability to articulate problems and possibilities, and actually be part of modifying robot applications.

1.2 Robots as Part of Digitalization

When the term robot is used in this paper, it refers to a humanoid-like machine that is autonomous or semi-autonomous different, from other artifacts that also include robotics and artificial intelligence – for example, Siri in iPhone or Google home assistant.

Discussing robots in care work must be viewed in the perspective of how technology is valued and distributed in this context. In Sweden there is a strong tradition of evaluating and distributing technologies to elderly people within the framework of assistive devices and occupational technologies. Technology becomes an extension of professional measures for the elderly in their role of patient and care receiver. Hence, robots become part of care concepts and welfare programs [9]. This is different from talking about robots as a three-part interaction between teachers, students and the robot [23]. In this context, the interaction is between the assistant nurse, the receiver of the elderly care and the robot as well as other applications connected to the robot – for example, sensors or alarms. The ability of older people to learn and interact with a robot is very much dependent on what healthcare professionals expect from them. Often the technology is designed so that it only allows limited actions such as eating or showering. Also, their use of technologies is monitored and they might be told how they should be used and not used. Learning is hardly on the agenda.

The consideration of constructivism as an idea for how learning and use develop in interaction between teachers, students and robots, is an aspect promoted in the literature [23]. This is of course possible to apply in care for the elderly as well but presupposes the study of this interaction in practice, which is not the purpose of this paper. In this paper, we are limited to the teachers' own experiences and reflections on robots in care for the elderly.

Having said that, robots in this paper is considered as part of the broad developments of digitalization which refers partly to a further extension of the use of IT where even more information is transformed into digital form, and partly to the increase in digital applications in our daily lives. An example of a technology in care for the elderly that is a further extension of the use of IT is personal emergency response systems (PERS), which have been shown to withstand several technology shifts and which are today being digitized as part of the third generation security alarm system. The robot is one in a range of new digital applications in care for the elderly including multimedia for interactive sessions, iPads, smartphones and other applications that facilitate communication, sensors in health monitoring systems, exoskeletons to support rehabilitation, digital locks and planning systems [24–27].

2 Method

Learning to use robots in care for the elderly as part of assistant nurse education is the research object in this paper. The four objectives are organized in a procedure with four phases in which these four dimensions are examined: expectations, course developments, robots in care practice and design care practice, see Table 1. The first phase concerning expectations involved 35 teachers in a seminar led by one researcher from KTH. Two issues were addressed, namely their own expectations of robot use in care for the elderly and the expectations from society as they are expressed in policy documents [21, 28].

The second phase concerning course development involved two experienced teachers engaged in teaching technology in two meetings with the KTH researcher. Existing teaching materials including books, digital platforms or other tools that they used were examined and evaluated in terms of their usefulness for teaching robotics or technology in general during assistant nursing education.

The third phase concerning robots in care practice involved the same 35 teachers as for the first phase in a seminar with a focus on the practice of inventing potential problems and possibilities.

The fourth phase involved 14 teachers who registered at the previous seminar. A design workshop was organized according to a procedure in which the group is initially presented with three alternative artifacts, in this case three robot applications. Thereafter, three steps follow in which the participants select one artifact from the following three choices; motivate and concretize the choice; modify the artifact out of their experiences of fitting technology into care work and what they understood as the needs of old people; and evaluate the result [16, 29].

Table 1. Project design with four phases, participants and methods.

Phases	Participants	Method
Expectations	35 Teachers	Seminar
Course developments	2 Teachers	Inventory
Robots in care practice	35 Teachers	Seminar
Designing care robots	14 Teachers	Design Workshop

2.1 Participants

The participants include teachers in assistant nursing education at three colleges in Stockholm. Invitations to participate were handled by the communicator at VO-COLLEGE. The teachers received an invitation and registered whether or not they were interested in participating. Meetings took place at KTH or VO-COLLEGE led by professor Britt Östlund, and for the design workshop, also by Björn Fischer PhD student at KTH.

In total, there were 49 participants in the project, distributed over two seminars, two course development meetings, and one design workshop planned over three years. Thirty-five persons participated in the seminars, fourteen persons in the design workshop, and three persons in course development. The vast majority, apart from two participants, was female.

2.2 Collection and Analysis of Data

At the seminars, the participants were divided into small groups where different issues were discussed. Each group was responsible for presenting its conclusions, which were written down on flipcharts. The KTH researcher who led the seminars collected the flipcharts, compiled the points and presented the conclusions at the following seminar or meeting. Participants could then confirm or reject the results.

During the inventory of teaching materials in the second phase, the researcher kept notes, which were then distributed to the two teachers for a joint result.

The design workshop was observed by a researcher, who compiled the participants' work into four steps according to the design workshop method used. The results were compared with other groups that participated in the same kind of design workshop and were published at a conference in Human Computer Interaction 2020 [16].

3 Result

3.1 Expectations

Teachers' own expectations and expectations from society articulated in policy documents concerning digitalization in care for the elderly in general and robots in specific were discussed. Questions were raised about working methods, leadership and the roles of the nurse and caregiver. Questions concerning robots specifically dealt with the difference between healthcare and self-care and who is responsible for the maintenance of technology. A third type of questions concerns any differences between receiving support from a robot and from a human. Conclusions from the groups' discussions were presented as follows:

The reports from the discussions include the following views and new questions:

a) The technological developments seem to create more systems than individual applications with increased complexity; issues of leadership and collaboration are becoming increasingly important.
b) When is it healthcare and when is it specialized self-care? Who is responsible for technical equipment in the spaces? Interstate support in the ambulances?
c) Is the patient / citizen / "person" as important in the system as with a real-life caregiver?

Speaking of changing professional roles, who are the experts on what happens at home?

d) Who implements the technology?
e) What about security?
f) Are we really interested in these changes? Some suggestions for why change is slow: individual resistance to change or tired of organizational changes, generations have different motives for change or non-change, no smooth contact with the IT-department or lack of resources.
g) How to support the use of technology by patients and caregivers?

3.2 Course Development

The first meeting on course development consisted of an inventory of what teaching material exists today. Two experienced teachers with a commitment to technology issues participated.

The results showed that the availability of teaching materials is unsatisfactory and the teachers may try to collect what they themselves consider relevant. They use research reports that may be inspiring for teachers but which are at too high a level for students; a book on medical technology that can serve as a reference book; and well-crafted chapters in nursing books that are relevant but quickly become obsolete. Dictionaries, for example, can change quickly. There are a couple of learning platforms, developed by individual entrepreneurs, useful for the teachers. There are also policies and political strategies here, but in such cases it is about visions rather than use. To summarize, the focus in accessible books is very much on medical equipment for hospitals and home healthcare, not on care work. The inventory shows a lack of content that contributes to achieving the goal of the education in order to give students the ability to actively contribute to developments and organize activities in a sustainable way.

The aim of the second meeting on course development was to formulate what teaching materials are needed to meet the technical development. The discussion began by stating that students' IT knowledge is greatly overestimated. They know Instagram and the smartphone but cannot handle computers, attach files, search for information as expected. Many students do not even collect their computer that they get from VO-COLLEGE. Students with a foreign background lack language support. "Digitalization is too big for us says the teachers, that is why we have to collaborate with KTH and others". Robots in particular easily lead us to think about movies and that we can easily talk to them. Important issues are the need for physical contact; to meet others; what can robots do; what are people good at?; what are robots good at?

The meeting concluded that the following is necessary for teaching about technology and technological developments:

a) For students: since documentation is increasing in importance, documents for practice with the language are desirable, especially for those with a foreign background.
b) For teachers, the need for competence development to keep up with developments

c) The need for uniform systems and templates to collaborate with and that are economically feasible.
d) Content to discuss the care profession of assistant nurses and the culture of the municipalities where they will work.
e) Technology related to the technological landscape in practice, what does it look like?
f) Patient safety, legislation and rules.

To summarize, the meeting concluded that teaching materials today considering digitalization should be about documentation, management and development, not only teaching the ability to use single applications, but also to improve the ability to manage critical perspectives and participate in developments. One way of participating is through design workshops, which was planned in the fourth phase of the project.

3.3 Robots in Care Practice

This seminar covered the potential and problems with robots in care practice. Some of the teachers participating in the seminar have first-hand experience of care work, while others had information they got from students on internships, which is a major part of the education. The seminar was introduced with the example of a care robot, the Giraff robot, which is a Swedish product tested as part of monitoring systems or as a single application for communication followed by discussions in small groups about what a robot can do in care work, what are the most important issues to be considered, and how robots will affect the assistant nurse professions.

The reports from the discussions includes the following views and new questions:

a) It is important to discuss how we can best make use of robots. What work tasks can be automated and what should not be automated?
b) It is important to be able to distinguish between the technology's "best-before date" and how it contributes to good care and nursing. Technology development is fast. The stuff may already be too old when the students finish their education.
c) It is important to know how people can influence and take responsibility for technology development, procurement and critical thinking in their professional roles.
d) It is important to consider how to exercise digital leadership.

3.4 Workshop: Designing Care Practice

The last gathering was a design workshop in which 14 teachers participated. The theme of the workshop was: Care robots, utopia or reality? How should care robots be designed? For the teachers this was an opportunity to experience what it is to participate in design, and what they can contribute with to make products usable in care work. The design workshop was not intended to design anything from the beginning, three existing products were presented, and they had to choose one of them. All three included a robot including a smartphone, a digital home assistant, and a humanoid robot. The design workshop was planned in four steps: creating alternatives to design; selecting alternatives; modifying and then evaluating alternative designs After choosing a product, they had to justify the choice, and modify and then evaluate the product in terms of size, style, functions, systems or anything relevant to home care.

All participants chose the humanoid robot but not as a single application. They preferred to connect the robot to the home assistant and the smartphone, which would give users more than one option. They suggested many modifications – for example, an automatic translator to support communication between assistant nurses and care receivers with different languages, using the robot like a dog for walks, storytelling and music books. They also added services, for example medicine reminders. The applications were recommended to be waterproof because there may be occasions when water is spilled. Several problems were discussed – for example, the robot's ability to recognize voices and where sensors should actually be fitted in the home and for what. The evaluation blended into the discussions around modifications.

To summarize, the most important point coming out of the workshop was the opportunity to discuss possibilities and potential risks with robots and other digital applications. Their role can be interpreted as them being consultants, by contributing with their own views or student views, but not primarily based on their own experiences from their work.

4 Discussion

In this paper, the question is one of how technology and robots can a take place in the training of assistant nurses. So far, the discussion has been about what trends are discernible, what is needed to make robots work in human contexts, and what challenges we can probably expect based on what robots are coming [1, 2, 30]. Some evaluations show that the developments are slow due to lack of knowledge about the context and eco-systems in which robots are supposed to be implemented [31]. Comparisons show, despite of high political ambitions, that the care sector is lagging behind other sectors when it comes to the uptake of robots [3–5].

The results from the seminars and the design workshop reported in this paper show that there is a long way from there being an interest in robots to actually contribute with suggestions on how they should be designed or used in order to be useful in care for the elderly. Design workshops seem to be a more useful method compared to seminars to generate suggestions for how robots can be better adapted to care work. The seminars brought up broader, though important, questions about care work and the role of assistant nurses and care receivers. It also takes time for them to understand that they themselves have something to contribute, despite their long experience of technology in care for the elderly. This may be a consequence of the notion that the development and design of robots only belong to engineers, and are nothing that healthcare professionals should have a part in, and that the distribution is controlled by specially trained people who are experts in what the elderly need and can handle.

The scarce contributions to defining problems and possibilities might be because the teachers themselves actually have less experience of work in care for the elderly or that they are not used to discussing technological aspects. Not least is it difficult to discuss technology or any routines and habits that actually work very well since they have a tendency to become invisible, which is a well-known methodological problem

[35]. New and problematic technology is easier to discuss since it is conspicuous and often creates unforeseen problems. Using design workshops as is reported in this paper is one way that seems to release experiences and unlock creativity.

The assumption that the development of a curriculum for assistant nurses requires input from those who know care work in practice or at least from students is not convincing. Perhaps a closer collaboration between engineering, design and teaching assistant nurses can do the job as well. The inventory of teaching materials proves that it is possible to go further when the questions are concretized. However, the results also show that using a more strict method might contribute to more ideas and comments, as was the case in the design workshop. Another suggestion is to collaborate in periods of internship with a follow-up in the classroom [18].

Questions raised about how robots affect the relationship between the caregiver and the care receiver and any differences between a robot and a human being are important for future considerations of what can or cannot be automated in care for the elderly. With regard to the organization of care for the elderly, available research shows a deteriorating work environment, but without including digitalization or technological development as factors affecting the work environment [33, 34]. Future research is necessary both to optimize the design of robots to be used in care for the elderly and to contextualize the use and evaluate the effects on the work environment and the content of the care provided. Bringing robot development closer to this environment can also help to avoid excess resources being allocated to faulty solutions and, in the worst case, reinforcing stereotypes about what care entails and the abilities and preferences of old people.

5 Conclusions

Integrating robots or any technology relevant in care work is still an open possibility making technological development more accurate and at the same time contributing to the development of care work as it raises broader issues such as leadership, ethics and roles. Increased responsibility for this development conveyed in the education of assistant nurses as well as other healthcare professions can help to broaden the view of technology, from being passive recipients of technology to being active co-creators.

Acknowledgements. The College of Care and Welfare, VO-COLLEGE, deserves thanks for contributing to the knowledge development of robotics in the education of assistant nurses. A special appreciation to Björn Fischer who assisted in the design workshops.

References

1. Frennert, S., Aminoff, H., Östlund, B.: Technological frames and care robots in eldercare. Int. J. Soc. Robot. (2020)
2. Royakkers, L., van Est, R.: A literature review on new robotics: automation from love to war. Int. J. Soc. Robot. **7**, 549–570 (2015)

3. Östlund, B., Malvezzi, M., Frennert, S., Funk, M., Gonsalez-Vargas, J., Moreno, J., Baur, K., Dimitris, A., Thorsteinsson, F., Cepeda-Alonso, A., Guillaume, F., Haufe, F., Di Pardo, M.: Uptake of Interactive Robots to promote health in Europe. What is the adoption potential and innovative maturity? INBOTS paper submitted to JN Social Robotics October 2020. In Review
4. Oxford The future of employment: How susceptible are jobs to computerization? Frey and Osborne, Oxford (2013)
5. McKinsey A future that works: Automation, Employment and productivity, McKinsey (2017)
6. Litinas, A., Alimisis, D.: Planning, implementation and evaluation of lab activities using robotic technology for teaching the phenomenon of motion. In: Ladias, A., Mikropoulos, A., Panagiotakopoulos, C., Paraskeva, F., Pintelas, P., Politis, P., Retalis, S., Sampson, D., Fachantidis, N., Chalkidis, A. (eds.) Proceedings of the 3rd Pan-Hellenic Conference "Integration and Use of ICT in Educational Process". HAICTE & University of Piraeus, Piraeus (2013). (in Greek)
7. YRGO, Polytechnic educations Gothenburg. https://yrgo.se/utbildningar/underskoterska-stodassistent-med-specialistkompetens-inom-valfardsteknologi/. Accessed 14 Oct 2020
8. Alimisis, D.: Educational robotics: open questions and new challenges. Themes Sci. Technol. Educ. **6**(1), 63–71 (2013)
9. The National Board of Health and Welfare: E-hälsa och välfärdsteknik I kommunerna 2020. Uppföljning av utvecklingen inom e-hälsa och välfärdsteknik I kommunerna [E-Health and welfare technology in the municipalities 2020. Follow-up of the development within e-health and welfare technology in the municipalities] (2020)
10. Governmental report: National Strategy for Life Science (2019)
11. The National Board of Health and Welfare: E-hälsa och välfärdsteknik i kommunerna [E-health and welfare technology in municipalities] (2019)
12. Hjalmarsson, M.: New technology in home help services — a tool for support or an instrument of subordination? Gender Work Organ. **16**(3), 368–384 (2009)
13. Östlund, B., Frennert, S.: How have user representations been sustained and recreated in the design of technologies between 1960 and 2018? In: Peine, A., Marshall, B., Martin, W. (eds.) Socio-Gerontechnology – Interdisciplinary Critical Studies of Age and Technology. Routledge, March 2021
14. Bratteteig, T., Wagner, I.: Disentangling Participation: Power and Decision-Making in Participatory Design. Springer, Cham (2014)
15. Fischer, B., Östlund, B.: Technology development with older people: the role of "unfettered design". In: Gao, Q., Zhou, J. (eds.) HCI 6th International Conference 2020. Human Aspects of IT for the Aged Population. Technologies, Design and User Experience, Copenhagen, pp. 18–33 (2020)
16. Östlund, B., Fischer, B., Marshall, B., Dalmer, N., Fernandez-Ardévol, M., Garcia-Santesmases, A., Lopez, D., Loos, E., Chang, F., Chen, X., Neven, L., Peine, A., Rosales, A., Kuoppamäki, S.: Using academic work places to involve older people in the design of digital applications. Presentation of a methodological framework to advance co-design in later life. In: Gao, Q., Zhou, J. (eds.) HCI 6th International Conference 2020. Human Aspects of IT for the Aged Population. Technologies, Design and User Experience, Copenhagen, pp. 45–58 (2020)
17. Peine, A., Faulkner, A., Jeager, B., Moors, E.: Science, technology and the 'grand challenge' of ageing—understanding the socio-material constitution of later life. Technol. Forecast. Soc. Change **93**, 1–9 (2015)
18. Mattsson, J., Eriksson, A., Östlund, B.: Creating learning opportunities for nursing students and medical engineering students. J. Res. Interprof. Pract. Educ. (JRIPE) **9**(1) (2019)

19. Julio, F., Chen, L., Bhutta, Z.A., Cohen, J., Crisp, N., Evans, T., Fineberg, H., Garcia, P., Ke, Y., Kelley, P., Kistnasamy, B., Meleis, A., Naylor, D., Pablos-Mendez, A., Reddy, S., Scrimshaw, S., Sepulveda, J., Serwadda, D., Zurayk, H.: Health professionals for a new century: transforming education to strengthen health systems in an interdependent world. Lancet **6**(9756), 1923–1958 (2010)
20. Statistics Sweden: Trends and forcasts 2017. Stockholm, Sweden: Statistics Sweden. U.S. Department of Labor, B. o. L. S. (2020). Occupational outlook handbook (2017). https://www.bls.gov/ooh/healthcare/home-health-aides-and-personal-care-aides.htm
21. European Commission 1994: Europe and the Global information Society (Bangemann report). www.epractice.eu/files/media/media_694.pdf. Tillgänglig 2012-07-24
22. En Nationell Strategi för Life Science. Government Offices: Ministry of Enterprise and innovation. Stockholm, artikelnummer N2019.06 (2019)
23. Catlin, D., Blamires, M.: The principles of educational robotic applications (ERA). A framework for understanding and developing educational robots and their activities. In: Clayson, J., Kalaš, I. (eds.) Constructionism 2010: Constructionist Approaches to Creative Learning, Thinking and Education: Lessons for the 21st Century: Proceedings for Constructionism 2010: The 12th EuroLogo Conference (2010)
24. Coradeshi, S., Cesta, A., Cortellessa, G., Gonzalez, J., Karlsson, L., Forsberg, A., Frennert, S., Furfari, F., Loutfi, A., Orlandini, A., Pecora, von Rump, S., Östlund, B.: GiraffPlus: a system for monitoring activities and physiological parameters and promoting social interaction for elderly. In: Human-Computer Systems Interaction. Backgrounds and Applications, Series Advances in Soft Engineering, vol. 3. Springer (2013)
25. Shibata, T.: Therapeutic seal robot as biofeedback medical device: qualitative and quantitative evaluations of robot therapy in dementia care. Proc. IEEE **100**(8), 2527–2538 (2012)
26. https://www.camanio.com/sv/products/bestic/
27. Robotics Care Inc.: https://roboticscare.com/robotics-care-poseidon/
28. Government Offices Sweden: a) SOU 1994, b) SOU 1995, c) SOU 2014:13. En digital agenda i människans tjänst – en ljusnande framtid kan bli vår. Delbetänkande av Digitaliseringskommissionen
29. Bratteteig, T., Wagner, I.: Disentangling participation. Power and decision-making in Participatory Design. Springer (2014)
30. Beasley, R.A.: Medical robots: current systems and research directions. J. Robot. (2012)
31. Van Aerschot, L., Parvaiainen, J.: Robots responding to care needs? A multitasking care robot pursued for 25 years, available products offer simple entertainment and instrumental assistance. Ethics Inf. Technol. **22**, 247–256 (2020)
32. Frennert, S.: Hitting a moving target: digital transformation and welfare technology in Swedish municipal eldercare. Disab. Rehabil. Assist. Technol. **16**, 103–111 (2019)
33. Rydenfält, C., Holgersson, C., Östlund, B., Arvidsson, I., Johansson, G., Persson, R. Picking low hanging fruit – a scoping review of work environment related interventions in the home care sector. Home health services quarterly. Accepted for publication (2021)
34. Rydenfält, C., Persson, R., Arvidsson, I., Holgersson, C., Johansson, G., Östlund, B., Persson, J..: Exploring local initiatives to improve the work environment – a qualitative survey in Swedish home care practice. Home healthcare management and practice, Accepted for publication (2021)
35. Berger, P., Luckmann, T.: The Social Construction of Reality. A Treatise in the Sociology of Knowledge. Penguin Books Ltd. (1991)
36. Frennert, S., Forsberg, A., Östlund, B.: Elderly people's perceptions of a telehealthcare system: relative advantage, compatibility, complexity and observability. J. Technol. Hum. Serv. **31**, 218–237 (2013)

New Studies and Methodological Issues in Educational Robotics

Educational Robotics Acceptance by Italian Teachers, Educators, Psychologists and Psychotherapists

Silvia Di Battista[1(✉)], Monica Pivetti[1], Brunilda Simaku[2], Gloria Beraldo[3,4], Emanuele Menegatti[3], and Michele Moro[3]

[1] Department of Human and Social Sciences, The University of Bergamo, 24100 Bergamo, Italy
{silvia.dibattista,monica.pivetti}@unibg.it
[2] Department of Psychological, Health and Territorial Sciences, The University of Chieti-Pescara, via dei Vestini 31, 66100 Chieti, Italy
[3] Department of Information Engineering, The University of Padova, 35124 Padua, Italy
gloria.beraldo@phd.unipd.it, {emanuele.menegatti, michele.moro}@unipd.it
[4] Institute of Cognitive Sciences and Technologies, National Research Council of Italy, Rome, Italy

Abstract. The exploration of acceptance of robotics by teachers, educators, psychologists and psychotherapists appears vital for the effective implementation of Educational Robotics (ER). This exploratory study evaluated acceptance of ER by Italian teachers, educators, psychologists and psychotherapists ($N = 88$). This research used a cross-sectional exploratory approach with a self-administered survey based on the Unified Theory of Acceptance and Use of Technology (UTAUT). The results indicated that the professionals' intention to use robotics in their profession was negatively related to anxiety evoked by robotics, but positively related to the following aspects: 1) positive attitudes about the use of robotics, 2) perceptions that there are factors in the environment that facilitate using robotics, 3) perceptions of ability of robots to be adaptive, 4) perceptions of robotics as enjoyable, 5) perceptions that robotics will improve the user's workplace performance, 6) positive perceptions of what other people think about robotics, 7) perceptions that robotics performs with integrity. Teachers, educators, psychologists and psychotherapists declared themselves to be quite favorable towards the use of robotics in their profession. These results showed that ER may be more integrated in educational contexts and for therapeutic and rehabilitation purposes.

Keywords: Educational robotics · Unified Theory of Acceptance and Use of Technology (UTAUT) · Acceptance of educational robotics

1 Introduction

The widely felt need for a school renewal at all levels also passes, and, according to some researchers and professionals, essentially, through the use of a 'good technology'. In the field of education, technology is 'good' if it is not reduced to a new skill to be acquired

M. Malvezzi et al. (Eds.): EDUROBOTICS 2021, SCI 982, pp. 167–178, 2021.
https://doi.org/10.1007/978-3-030-77022-8_15

but is rather an incentive to learning tool, a cognitive engine that stimulates interest and gives pleasure to the user. Among the technologies that can be classified like this, there is robotics: the fact that it has been successfully applied in recent years in education is confirmed by the birth of a new expression, "Educational Robotics", that has been coined and currently used in literature with a meaning rich in value and contents.

Research reveals that robotics can be a valuable tool for the educational field in the learning process of average students [1–3] and of students with special needs (SNs) [4]. As for students with special needs, Pivetti and colleagues [4] recently found that ER emerged as a valuable instrument for students with neurodevelopmental disorders (from 3 to 19 years of age). Furthermore, psychologists and educators have shown the positive psychological impact of using robotic construction kits on the development of cognitive and social skills [5].

The study of all those factors that influence behavioral intention and the effective use of educational robotics seems very important because owning the equipment and using it are not the same. Previous studies [6] have found that most teachers have positive attitudes towards ER agreeing to consider it as a powerful tool for promoting several skills for students and children with SNs. However, other studies also highlighted that there is still some suspicion towards the use of robots in the fields of education and care of people [7, 8]. This suspicion could have a negative impact in the integration of ER [7, 8]. For instance, a European survey (European Commission, 2017) [9] showed that only 26% of respondents were comfortable "with having a robot to provide services and companionship when infirm or elderly" and only 35% declared themselves to be comfortable in having a robot assist them at work. Furthermore, Denmark was the only country where at least one in ten respondents use a robot at work (14%), compared to 6% in Italy.

Heerink and colleagues [8] argued that, even in the face of scientific evidence, robots are often not implemented in the fields of education and care of people because of factors like stigmatization, perceptions of (non)adaptability of robotics or negative social influences. Therefore, the authors discussed that it is not only vital to understand the user acceptance of robotics in order to provide insight in the probability of using robotics, but also to understand the factors underlying acceptance propensities. Other researchers argued that understanding people's attitudes and acceptance of robotics is a prerequisite for predicting their intention to use ER and for implementing effective interventions in ER [10]. For instance, factors that influence user acceptance and adoption of ER can inform us about the priorities for fundraising, training needs or research programs [10]. Furthermore, we believe that ER is of extreme interest not only for educators and teachers but also for psychologists and psychotherapists as a valuable tool to support and develop cognitive and social skills, problem solving, critical thinking, innovative ideas, and team working skills among others [11]. Psychologists and psychotherapists often work with teachers and educators for socio-therapeutic purposes, in particular for the evaluation and the intervention in case of special needs students. Therefore, teachers, educators and mental health professionals could benefit from knowing about the wide range of possible applications and implications of ER. Furthermore, these targets of professionals' attitudes towards ER are fundamental in predicting the intention to use robotics in work. Consistent with these premises, the present study aims to expand the line of research into the acceptance of ER by investigating Italian teachers, educators and psychologists/psychotherapists' attitudes. In Italy, the terms "educators" and "teachers" have several and overlapping meanings. In this paper, we intend educators (in Italian: "educatori dei servizi educativi per l'infanzia") as the professionals working in nursery schools (Decree n° 378/2018, Law n° 65/2017). Teachers are defined as the professionals teaching in kindergartens, primary schools, middle and secondary schools (Contratto Collettivo Nazionale del Comparto Scuola – CCNL, 2006/2009)[1]. In Italy, the profession of psychologist comprises the use of cognitive and operative instruments for the prevention of problems, for diagnosis, and for activities relating to psychological rehabilitation and support of individuals, groups, social bodies and communities, including experimental, research and teaching activities within these areas. The practice of psychotherapy is subject to special professional training (see the National Board of Italian Psychologists, reference number: XV/E/58452/95).

The theoretical background of this study is based on the assumptions of the Unified Theory of Acceptance and Use of Technology (UTAUT) [12].

1.1 The Unified Theory of Acceptance and Use of Technology (UTAUT)

A user's acceptance of technology is defined as "the demonstrable willingness within a user group to employ technology for the tasks it is designed to support" [13]. The first studies on technology acceptance were based on the Technology Acceptance Model (TAM) by Davis in 1989 [14] that was, in turn, grounded on the research of social psychologists Ajzen and Fishbein [15]. The TAM theory specified that the user's perceptions of usefulness and ease of use of a certain technology impact behavioural intention to use it. The theory also assumed that this behavioural intention is predicting the actual use of a specific system. Venkatesh and colleagues [12] unified several models and findings on technology acceptance in the so-called Unified Theory of Acceptance and Use of Technology (UTAUT) that was tested for many different types of technology [16]. This model has been recently extended for the study of acceptance of robotics in the fields of education and care of people [7, 8]. In Conti and colleagues [7] and Heerink and colleagues [8], the UTAUT model comprises 13 dimensions measuring technology acceptance. These dimensions are represented in a questionnaire by a group of questions or statements that can be evaluated. The 13 dimensions are as follows: • Anxiety (ANX), comprising an unpleasant emotional state or condition which is characterized by subjective feelings of tension, apprehension, and worry [17] evoked when using a specific technology; • Attitude (ATT), comprising individuals' positive or negative evaluation about the use of a specific technology [15]; • Facilitating Conditions (FC), comprising all the perceived factors in the environment that facilitate using a specific technology [7, 8]; • Intention To Use (ITU), describing the user's intention to use a specific technology over a long period of time [7, 8]; • Perceived Adaptability (PA), referring to the perceived ability of a specific technology to be adaptive to the changing needs of the

[1] In Italy, according to the law 205/2017, the roles of socio-pedagogical professional educator and the pedagogic professional were also defined. These professionals work in the educational and pedagogical field (not only in schools), in relation to any activity carried out in a formal and informal way, towards people of all ages, from a perspective of personal and social growth.

user [7, 8]; • Perceived Enjoyment (PENJ) representing the extent to which the activity of using a specific technology is perceived to be enjoyable in its own right, aside from any performance consequences resulting from system usage [14, 18]; • Perceived Ease Of Use (PEOU) representing the degree to which the user believes that using a specific technology would be free of effort [7, 8]; • Perceived Sociability (PS) comprising the perceived ability of a specific technology to perform sociable behaviours [7, 8]; • Perceived Usefulness (PU) comprising the user's perception to the extent that a specific technology will improve the workplace performance [14, 18]; • Social Influence (SI) comprising the user's perception of how people, who are important to him/her, think about him/her using the system [7, 8]; • Social Presence (SP) that is a perceived personal experience of sensing a social entity when interacting with a specific technology [7, 8]; • Trust (TRST) comprising the user's perception that a specific technology performs with integrity and reliability [7, 8]; • Use (USE) comprising the actual use of the system over a longer period of time [7, 8].

In the study of Heerink and colleagues [8] the questionnaire based on the UTAUT was administered to 30 participants, recruited both by eldercare personnel and by students, in order to explore the acceptance of assistive social robots (including in this category all those robots that are in any way assistive to elderly adults and that are socially interactive). Results showed that the intention to use the assistive social robots predicted the actual use and it was significantly determined by two factors: 1) a perception of ease of use technologies (e.g.: "I find the robot easy to use"); and 2) positive attitudes (e.g.: "The robot would make my life more interesting"). Results also showed that the perception of ease of use was influenced by the perceived usefulness of technology (e.g.: "I think the robot is useful to me") and this latter was influenced by the perception of adaptiveness (e.g.: "I think the robot can be adaptive to what I need"). In the Italian context, after a brief interactive demonstration of the NAO robot capabilities in a class, Conti and colleagues [7] administered the UTAUT questionnaire to 80 participants (i.e.: 25 educational and rehabilitation care practitioners and 55 university students in psychology or education sciences) with no previous experience of robotics. Their results showed the applicability of the UTAUT model in the context of education and care of children, and suggested a positive attitude towards the use of a robot. Conti and colleagues [19] also surveyed Italian and British-English psychology students after an interactive demonstration using a humanoid robot to evaluate their social robot's acceptance and their intention to use a robot as an instrument for future clinical practice. Results showed that Italian psychologists were more positive toward the perceived usefulness and intention to use the robot in psychological practice compared to English psychologists. Furthermore, all respondents felt they did not have the necessary abilities to make good use of robotics.

2 The Current Study

The exploration of acceptance of robotics by teachers, educators and psychologists/psychotherapists appears vital for the effective implementation of ER [7, 19]. While user acceptance and adoption behavior have been studied for teachers and educators working in schools much less attention has been paid to the study of factors associated with the acceptance of ER for psychologists or psychotherapists. This result could reflect the difficulty of Italian psychologists/psychotherapists in using educational robotics and knowing its potential, considering the marginal role that in Italy these professionals have in educational contexts where ER activities are mainly integrated [20, 21]. In general, in research it is still unclear to what extent Italian teachers, educators and psychologists/psychotherapists are adopting and accepting ER for their work. In Italy, with the introduction of the National Digital School Plan in 2015, many schools have included ER, recognizing it as motivating, engaging, and a great learning tool particularly when special needs are concerned. In the Italian schools, increasing attention is paid to the topic of Special Educational Needs (e.g., law n° 170/2010). Furthermore, individualized plans for students with special education needs were established (i.e. Piano Educativo Individualizzato – PEI). These plans are developed by learning support teachers in collaboration with students' parents, with all school staff and with socio-health personnel (e.g., mental health professionals). PEI is based on the International Classification of Functioning, Disability and Health (ICF) of the World Health Organization. The ICF is defined as a classification system that aims to describe, not only diseases or disabilities, but also the state of health of individuals in a positive key (functioning and health) and in relation to their environment (social, family, work) in order to grasp the difficulties that in the socio-cultural context of reference can cause disabilities [22]. The personalized teaching defined in PEI is substantiated through the use of a variety of teaching methodologies and strategies, rehabilitative interventions (e.g., psychological interventions), didactic tools and assistive technology, in order to promote educational potential and success in each student. Technologies find a specific location also within ICF, described as learning-interaction tools that can represent facilitation/compensation instruments in the performance of normal activities and in the performance of learning tasks.

Given that there is still some skepticism towards the use of robots in the fields of education and care of people [7, 19] and data on psychologists' and psychotherapists' acceptance of ER are scarce, this exploratory study aimed to evaluate user acceptance of ER by Italian teachers, educators and psychologists/psychotherapists. Specifically, this study aimed to use the UTAUT model to examine factors that affect acceptance behavior of ER in the workplace by Italian teachers, educators and psychologists/psychotherapists. In this study, we used a cross-sectional exploratory approach with a self-administered survey.

We expected that, among teachers, educators and psychologist/psychotherapists, the intention to use a robot in one's future profession would be negatively related to anxiety but positively related to 1) positive attitudes towards robotics, 2) the facilitating conditions (such as the knowledge of robotics), 3) the perceived adaptability of robotics in their own work, 4) the perceived enjoyment in using a robot, 5) the perceived usefulness of robotics for their profession, 6) social influence and 7) perceptions of trust in robotics.

3 Method and Materials

3.1 Participants and Procedures

Participants were recruited by snowball sampling. A total of 111 questionnaires were administered online. Three participants did not give their consent to the participation and so were entirely excluded from the data collection. Twenty-one participants were excluded from the analyses because they failed at the attention check questions (see Materials section below). They were re-directed to the debriefing section. The final sample consisted of 88 experienced and future professionals (24 psychologists/psychotherapists, 27.3%; 64 educators/teachers), 73 females and 15 males (females = 83%) aged from 23 to 68 years ($M = 43.51$ years, $SD = 13.69$). The sample was composed of professionals with different specializations (e.g., math teachers, psychodynamic psychotherapists among others). The majority of the participants were highly educated with 80.7% having a university degree or a post-graduate degree or a PhD qualification ($n = 71$) and 19.3% having completed secondary school ($n = 17$). According to their working experience, 51.4% of them had been working for at least 17 years (variable ranging from 1 up to 45 years; $M = 18.42$; $SD = 13.55$). Fourteen participants were pre-service professionals (15.9%). Participants originated from regions of the Centre ($n = 28$; 31.8%), the South or the Islands ($n = 24$; 27.3%), and the North ($n = 36$; 40.9%) of Italy.

Twenty-three participants had never heard about ER (26.1%), 18 participants (20.5%) declared that they had few notions about it, 21 participants (23.9%) stated they knew a little about ER, and 26 participants (29.5%) reported they knew it well and very well. Thirty-six participants (40.9%) had never attended a course or a workshop about ER. As for previous experience with ER in their work environment, 53.4% of professionals had been using ER in their work ("few times" $n = 6$; "sometimes" $n = 11$; "many times" $n = 24$; "almost always" $n = 6$), whereas 46.6% of them ($n = 41$) had never used it (comprising all participants who were psychologists/psychotherapists). The questionnaire was implemented using the Qualtrics.com form. The survey was open from October 2019 to October 2020. After reading a description of the study and indicating their willingness to participate in an informed consent form, each participant read a script introducing ER and then they evaluated it (see Materials section below). Respondents read a debriefing statement at the end of their contribution. No remuneration was offered for participation. The research complied with the Ethics Code of the Italian Psychology Association (Associazione Italiana di Psicologia) [23], and was conducted in accordance with WMA-Declaration of Helsinki (1964/2013) [24]. As no Institutional Review Board for Psychology research was available from the affiliations of the social-psychology researchers involved in the study (i.e. University of Chieti-Pescara, Chieti, Italy and University of Bergamo, Bergamo, Italy), no request for approval was submitted.

3.2 Materials and Measures

The Script. After participants indicated their willingness to participate in the study in a complete informed consent form, they were invited to read a scenario describing educational robotics[2]. This description was inspired by the relevant literature on ER [1–3] and revised by one researcher in the ER field. After reading the script, participants answered two attention-checking questions (i.e.: "*Educational Robotics is suitable for students of all ages;*" "*Educational Robotics is suitable for students with special needs - e.g. autism spectrum disorders;*" answers: true or false).

[2] Supplementary materials file: https://mfr.osf.io/render?url=https://osf.io/m76kb/?direct%26mode=render%26action=download%26mode=render.

The UTAUT Questionnaire: We administered a short version of the UTAUT questionnaire (translated in Italian and back-translated in Conti et al. [7]; permission to use the scale was requested from the authors). The participants were asked to indicate their level of agreement to 25 statements on a five-point Likert scale (from 1 = totally disagree; up to 5 = totally agree). Participants first read: "*Take a few minutes and try to imagine using Robotics in your work. What is your first impression?*" They then rated the items corresponding to UTAUT 9 dimensions:

1. Anxiety (4 items; e.g.: "*If I were to use a robot in my work, I would be afraid of making mistakes;*" $\alpha = .84$).
2. Attitude (3 items; e.g.: "*It's good to make use of robots in my work*"; $\alpha = .98$).
3. Facilitating conditions (2 items; e.g.: "*I know enough about robots to make good use of it;*" $\alpha = .63$).
4. Intention to use (3 items; e.g.: "*I am certain to use robotics in my work in the near future*"; $\alpha = .92$).
5. Perceived adaptability (3 items; e.g.: "*I think robots can be adaptive to what I need in my work;*"$\alpha = .62$).
6. Perceived enjoyment (3 items; e.g.: "*I find robots enjoyable;*" $\alpha = .78$).
7. Perceived usefulness (3 items; e.g.: "*I think robots are useful to me in my work*"; $\alpha = .89$).
8. Social influence (2 items; e.g.: "*I think it would give a good impression if I were to use robots in my work*"; $\alpha = .76$).
9. Trust (2 items; e.g.: "*I would trust robots if they gave me advice.*" $\alpha = .62$).

Experiences with Educational Robotics. Participants then indicated their previous experiences with ER answering the following questions: "*Are you familiar with Educational Robotics*?" (responses on a five-point Likert scale from 1 = not at all; up to 5 = very much); "*Have you ever attended a workshop - or a lesson - on Educational Robotics*?" (responses on 5 point Likert scale from 1 = never; up to 5 = very often); "*Have you ever used Educational Robotics in your work*?" (responses on a five-point Likert scale from 1 = no, never; up to 5 = yes, very often).

Socio-Demographic Questions. We also measured the self-reported participants' gender (1 = male; 2 = female; 3 = other), age, region of origin in Italy (1 = Centre; 2 = Northern Italy; 3 = Southern Italy and the Islands), educational level, professional role (1 = educator/teacher or 2 = psychologist/psychotherapist), work experience in their own specialization (1 = no; 2 = yes) and years of work experience in their own specialization.

4 Results

Table 1 Shows the mean and standard deviations among all the variables and the correlations between all measures investigated in the study. The analyses indicated that almost all these measures were related.

Table 1. Means (standard deviation) and zero-order correlations among variables

	Means (SD)	1	2	3	4	5	6	7	8
1.ANX	M = 1.76 (0.81)	1							
2. ATT	M = 4.19 (0.94)	-.35**	1						
3. FC	M = 3.02 (1.02)	-.25*	.38**	1					
4. ITU	M = 3.73 (1.13)	-.43**	.51**	.75**	1				
5. PA	M = 3.13 (0.79)	-.17	.38**	.41**	.45**	1			
6. PENJ	M = 4.25 (0.83)	-.28**	.14	.68**	.46**	.48**	1		
7. PU	M = 3.72 (0.92)	-.24*	.43**	.71**	.64**	.62**	.59**	1	
8. SI	M = 3.45 (0.92)	-.12	.39**	.53**	.45**	.55**	.55**	.71**	1
9.TRST	M = 3.52 (0.94)	-.16	.29**	.53**	.42**	.54**	.55**	.65**	.59**

Note. ANX: Anxiety; ATT: Attitude; FC: Facilitating Conditions; ITU: Intention to Use; PA: Perceived Adaptability; PENJ: Perceived Enjoinment; PU: Perceived Usefulness; SI: Social Influence; TRST: Trust. * = $p < .05$; ** = $p < .01$

Significant positive relationships were found between facilitating conditions (such as the knowledge of robotics) and 1) positive attitudes, 2) intentions to use, 3) perceived adaptability, 4) perceived utility, 5) social influence and 6) trust. Positive attitudes towards robotics were also positively related to 1) intentions to use, 2) perceived adaptability, 3) perceived enjoyment, 4) perceived utility, 5) social influence and 6) trust. Furthermore, intention to use robotics was related with 1) perceived adaptability, 2) perceived enjoyment, 3) perceived utility, 4) social influence and 5) trust. Perceived adaptability, perceived enjoinment, perceived utility, social influence and trust were positively correlated to each other. Finally, anxiety was negatively related to facilitating conditions, positive attitudes, intention to use, perceived enjoyment and perceived usefulness of robotics in their own work. Furthermore, Fig. 1 shows the mean among all the variables for each subgroup (i.e. teachers, educators and psychologists/psychotherapists).

To test whether the two subgroups of participants statistically differed in terms of study variables, a multivariate analysis of variance (MANOVA) was run with two subgroups as independent variable and all the study variables as dependent ones. The multivariate effect of subgroup was significant ($Wilks's\ \Lambda = 0.71$; $F\ (9, 78) = 3.70$; $p = .001$; $\eta^2_p = .30$). The univariate analyses showed significant differences between the two

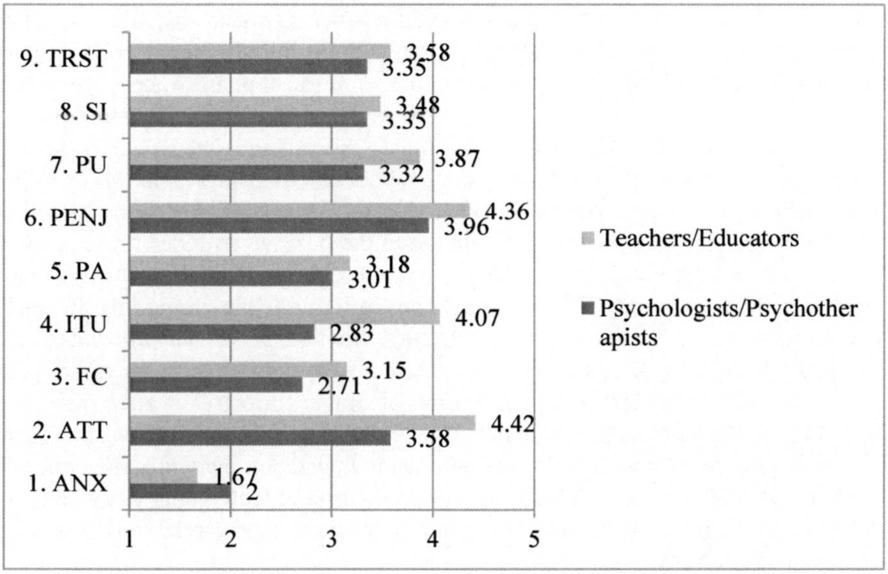

Fig. 1. Means of the examined variables per subgroup

groups. Specifically, teachers and educators showed higher positive attitudes towards robotics (F (1, 88) = 16.24; $p < .001$; $\eta^2_p = .16$), higher intentions to use robotics (F (1, 88) = 26.93; $p < .001$; $\eta^2_p = .24$), higher perceived enjoinment of robotics (F (1, 88) = 4.37; $p = .04$; $\eta^2_p = .05$), and higher perceived usefulness of robotics in their own work (F (1, 88) = 6.68; $p = .011$; $\eta^2_p = .07$) as compared to psychologists/psychotherapists. No other significant differences were found (all $ps > .05$).

5 Discussions

This research indicated that, for teachers, educators and psychologists/psychotherapists, the intention to use robotics in future professional work was negatively related to anxiety evoked by robotics but positively related to the individuals' 1) positive attitudes about the use of robotics, 2) perceptions that there are factors in the environment that facilitate using robotics (e.g. the knowledge of robotics), 3) perceptions of the ability of robotics to be adaptive to the changing needs of the users in their working environments, 4) perceptions of robotics as enjoyable in its own right, aside from any performance consequences resulting from system usage, 5) perceptions of utility or that robotics will improve the user's workplace performance, 6) positive perceptions of what other people think about robotics (that is the positive social influence), 7) perceptions that robotics performs with integrity and reliability. The first remark is dedicated to the evident, even not tremendously big, difference in perception between the group of teachers, educators and the 'therapists'. This is not surprising because this latter group has less opportunity to fully exploit robotics within their specialization and the literature regarding opportunities within the classroom by special needs support teachers is not so vast. This aspect could also be motivated by the difficulties associated with the introduction of robotics in clinical trials, from the ethical considerations to the recruitment phase where parents have to agree on the participation in the study as well as the involvement of qualified operators [25–27]. Teachers, educators and psychologists/psychotherapists declared themselves as being quite favorable towards the use of robotics in their profession. As for the differences between the two groups, teachers/educators showed higher scores as compared to psychologists/psychotherapists in relation to the positive attitudes towards robotics, intentions to use robotics, perceived enjoyment of robotics, and perceived usefulness of robotics in their own work. The differences are particularly high regarding the *intention to use* and *attitudes*. These two dimensions, together with *perceived enjoyment*, are also the ones with more relevant positive scores, but in general results are rather positive which confirm that ER has broken the barrier of simple curiosity towards a serious consideration of it as a usable learning tool. The *anxiety* negative aspect shows low numbers but *facilitating factors* does not show an outstanding score: evidently the general perception is still that the school system has some inertia in welcoming not widely tested technologies such as ER, which was further confirmed by the relatively low score of perceived adaptability.

In general, these study results show that ER could be more integrated in educational contexts and for therapeutic and rehabilitation purposes. Further studies should include larger and more balanced samples. Furthermore, future studies could explore teachers' opinions concerning the use of ER for engagement given the importance of this topic for psychological field [28] and inspect what other factors in the teaching and social context might influence the attitudes and use of robots. For instance, the focus on curriculum and high stakes testing may also affect the use of the robot in schools adversely [29–31]. Furthermore, the role of gender of the participants and/or sex of the robot could be investigated in terms of positive attitudes towards the use of robots in educational and mental health contexts [32].

Acknowledgments. We are grateful to Dr. Alba Di Primio for her comments and literature suggestions on the Italian education system. We also thank the reviewers for their efforts and much appreciated comments.

References

1. Benitti, F.B.V.: Exploring the educational potential of robotics in schools: a systematic review. Comput. Educ. **58**, 978–988 (2012). https://doi.org/10.1016/j.compedu.2011.10.006
2. Mubin, O., Stevens, C.J., Shahid, S., Al Mahmud, A., Dong, J.J.: A review of the applicability of robots in education. J. Technol. Educ. Learn. 1.209–0015: 13 (2013). https://doi.org/10.2316/Journal.209.2013.1.209-0015
3. Xia, L., Zhong, B.: A systematic review on teaching and learning robotics content knowledge in K-12. Comput. Educ. **127**, 267–282 (2018). https://doi.org/10.1016/j.compedu.2018.09.007
4. Pivetti, M., Di Battista, S., Agatolio, F., Simaku, B., Moro, M., Menegatti, E.: Educational robotics for children with neurodevelopmental disorders: a systematic review. Heliyon **6**, e05160 (2020). https://doi.org/10.1016/j.heliyon.2020.e05160

5. Caci, B., D'Amico, A., Cardac, M.: New frontiers for psychology and education: Robotics. Psychol. Rep. **94**(3_suppl), 1372–1374 (2004). https://doi.org/10.2466/pr0.94.3c.1372-1374
6. Agatolio, F., Pivetti, M., Di Battista, S., Menegatti, E., Moro, M.: Training course in educational robotics for learning support teachers. In: International Conference EduRobotics, pp. 43–57. Springer, Cham (2016). https://doi.org/10.1007/978-3-319-55553-9_4
7. Conti, D., Di Nuovo, A., Cirasa, C., Di Nuovo, S.: Robots in education and care of children with developmental disabilities: a study on acceptance by experienced and future professionals. Int. J. Soc. Robot. **9**, 51–62 (2017). https://doi.org/10.1007/s12369-016-0359-6
8. Heerink, M., Krose, B., Evers, V., Wielinga, B.: Measuring acceptance of an assistive social robot: a suggested toolkit. In: 18th IEEE International Symposium on Robot and Human Interactive Communication. IEEE Roman, pp. 528–533 (2009). https://doi.org/10.1109/roman.2009.5326320
9. European Commission: Special Eurobarometer 460: Attitudes towards the impact of digitisation and automation on daily life. Report, European Union (2009). https://doi.org/10.2759/835661
10. Liu, L.M., Cruz, A., Rios Rincon, A., Buttar, V., Ranson, Q., Goertzen, D.: What factors determine therapists' acceptance of new technologies for rehabilitation – a study using the unified theory of acceptance and use of technology (UTAUT). Disab. Rehabil. **37**, 447–455 (2014). https://doi.org/10.3109/09638288.2014.923529
11. Gaudiello, I., Zibetti, E.: Learning Robotics, With Robotics, by Robotics: Educational Robotics. Wiley. ISBN-10: 1786300990 (2016)
12. Venkatesh, V., Morris, M.G., Davis, G.B., Davis, F.D.: User acceptance of information technology: toward a unified view. Trimestrale MIS **27**, 425–478 (2003). https://doi.org/10.2307/30036540
13. Dillon, A.: User acceptance of information technology. In: Karwowski, W. (ed.) Encyclopedia of Human Factors and Ergonomics. Taylor and Francis, London (2001)
14. Davis, F.D.: Perceived usefulness, perceived ease of use, and user acceptance of information technology. MIS Q. **13**, 319–340 (1989). https://doi.org/10.2307/249008
15. Fishbein, M., Ajzen, I.: Belief, Attitude, Intention, and Behavior: An Introduction to Theory and Research. Addison-Wesley, Reading. ISBN: 0201020890 (1975)
16. Williams, M.D., Rana, N.P., Dwivedi, Y.K.: The Unified theory of acceptance and use of technology (UTAUT): a literature review. J. Enterp. Inf. Manage. **28**, 443–488 (2015). https://doi.org/10.1108/JEIM-09-2014-0088
17. Spielberger, C.D.: Conceptual and methodological issues in anxiety research. In: Spielberger, C.D., Sarason, I.G. (eds.) Anxiety: Current Trends in Theory and Research, vol. 2, pp. 481–493. Academic Press, New York (1972)
18. Davis, F.D., Bagozzi, R.P., Warshaw, P.R.: Extrinsic and intrinsic motivation to use computers in the workplace. J. Appl. Soc. Psychol. **22**, 1111–1132 (1992). https://doi.org/10.1111/j.1559-1816.1992.tb00945.x
19. Conti, D., Cattani, A., Di Nuovo, S., Di Nuovo, A.: Are future psychologists willing to accept and use a humanoid robot in their practice? Italian and English students' perspective. Front. Psychol. **10**, 2138 (2019). https://doi.org/10.3389/fpsyg.2019.02138
20. European Federation of Psychologist Associations [EFPA] Education, Training, Professional Profile and Service of Psychologists in the European Educational System (2010). www.efpa.eu
21. Matteucci, M.C., Chiesa, R., Albanesi, C.: Progettare un servizio di psicologia scolastica a partire dai bisogni della scuola: un progetto partecipato in un Istituto Comprensivo. Ricercazione **9**, 109–122 (2017)

22. Ferri, D.: Inclusive education in Italy: a legal appraisal 10 year after the signature of the UN convention on the rights of persons with disabilities. Ricerche di Pedagogia e Didattica–J. Theories Res. Educ. **12**, 1–22 (2017). ISSN: ISSN-1970-222123
23. Associazione Italiana di Psicologia, A.I.P. Code of Ethics of the Italian Psychological Association (2015). https://www.aipass.org
24. Declaration of World Medical Association (WMA) of Helsinki 1964–2013. https://www.wma.net
25. Papert, S.: The Children's Machine: Rethinking School in the Age of the Computer. BasicBooks, New York (1993)
26. Allmark, P.: The Ethics of Research with Children. (Ethical Issues in Research). Nurse Res. **10**, 7–20 (2002). https://doi.org/10.7748/nr2003.01.10.2.7.c5885
27. Beraldo, G., Menegatti, M., de Tommasi, V., Mancin, R., Benini, F.: A preliminary investigation of using humanoid social robots as non-pharmacological techniques with children. In: 2019 IEEE International Conference on Advanced Robotics and its Social Impacts (ARSO), Beijing, China, pp. 393–400 (2019). https://doi.org/10.1109/ARSO46408.2019.8948760
28. Berti, C., Di Battista, S.: Framing student engagement through perception of justice: the role of trust in authorities in educational contexts. In: Newley, R.J. (ed.) Classrooms: Management Effectiveness and Challenges, pp. 59–75, Nova Science Publishers, New York (2011)
29. Black, P., Wiliam, D.: Inside the black box: raising standards through classroom assessment. Phi Delta Kappan **92**, 81–90 (2010). https://doi.org/10.1177/003172171009200119
30. Catlin, D.: Using peer assessment with educational robots. In: International Conference on Web-Based Learning, pp. 57–65. Springer, Cham, August 2014. https://doi.org/10.1007/978-3-319-13296-9_6
31. Catlin, D.: Using learning intentions with educational robots. In: Constructionism 2016, Bangkok, Thailand, pp. 158–166 (2016)
32. Beraldo, G., Di Battista, S., Badaloni, S., Menegatti, E., Pivetti, M.: Sex differences in expectations and perception of a social robot. In: 2018 IEEE Workshop on Advanced Robotics and Its Social Impacts (ARSO), pp. 38–43. Yverdon-les-Bains, Switzerland (2018). https://doi.org/10.1109/ARSO.2018.8625826

Teachers' Reasons to Join a Community About Educational Robotics and STEAM: A Swiss Experience

Lucio Negrini[1(✉)], Sophia Reyes Mury[2], Dio Moonnee[3], Paolo Rossetti[4], and Melissa Skweres[2]

[1] Department of Education and Learning, University of Applied Sciences and Arts of Southern Switzerland (SUPSI), Piazza San Francesco 19, 6600 Locarno, Switzerland
lucio.negrini@supsi.ch, info@roteco.ch

[2] Centre for Learning Science LEARN, Swiss Federal Institute of Technology of Lausanne (EPFL), Route Cantonale, 1015 Lausanne, Switzerland

[3] mint&pepper, Wyss Zurich Project, Swiss Federal Institute of Technology of Zurich (ETHZ), University of Zurich (UZH), Weinbergstrasse 35, 8092 Zurich, Switzerland

[4] University of the Italian Switzerland (USI), Via Buffi 13, 6900 Lugano, Switzerland

Abstract. Today we live in a digital society that requires the acquisition of new skills related to computer science, such as computational thinking or coding skills, as well as cross-curricular skills, such as communication, collaboration and creativity. A possible tool that allows fostering these skills in schools is educational robotics. However, the question is how to bring educational robotics into schools? Indeed, only a few teachers are conducting educational robotics activities in their classrooms. This is where Roteco comes in. We aim to create a community of teachers interested in educational robotics and to offer training in this field in order to increase the presence of robotics in the classroom. As part of the project, a teacher training concept, a collaboration platform and various teaching resources were developed. This paper presents the project and the results of a survey with 87 teachers about the reasons why they joined the Roteco community and how they contribute to its growth. The main findings show that teachers joined Roteco to remain informed about robotics, to be inspired by other teachers and to get didactic resources. Only a few of them also joined in order to communicate with other teachers or to share their own materials. In order to create a vibrant community is however important to motivate teachers to be more active.

Keywords: Teacher training · Community · Educational robotics · STEAM · Computational thinking · Coding

1 Introduction

The integration of educational robotics, computer science (CS), and coding in schools is at the forefront of numerous initiatives in Switzerland and worldwide [3, 4]. Educational robotics is namely seen as a valuable tool to foster the acquisition of key competences

M. Malvezzi et al. (Eds.): EDUROBOTICS 2021, SCI 982, pp. 179–189, 2021.
https://doi.org/10.1007/978-3-030-77022-8_16

such as computational thinking, coding and transversal skills like communication, collaboration and creativity [2, 14], as well as to introduce pupils to science, technology, mathematics, arts and engineering (STEAM) [15]. Despite the increasing importance of educational robotics, only a few "pioneers" teachers are implementing it into classrooms [5]. In order to increase the number of teachers using robotics, there is a need for teacher trainings, support for "followers" through validated didactic activities and the experiences of other colleagues.

De Koster, Kuiper, and Volman [6] point out that the integration of technologies in schools is a complex process involving school policies, teacher attitudes, training, and a change in didactics and teaching practices. Another fundamental element in the successful integration of technologies in schools lies in the professionalism of the teaching staff, who are requested to partially modify their practices through continuous education [9, 16].

Different studies have shown that a teacher community can help to foster the collaboration among teachers through the sharing of materials or the common work on didactic activities as well as increment their specific knowledge learning through peers and change their attitudes and practices [7, 10, 17].

In this context, the Swiss Federal Institute of Technology of Lausanne (EPFL), the University of Applied Sciences and Arts of Southern Switzerland (SUPSI) and the Wyss Institute of the Swiss Federal Institute of Technology of Zurich (ETHZ) and of the University of Zurich, in partnership with the universities of teacher education of the Swiss cantons Vaud, St. Gallen, Lucerne, Schwyz and Valais started the Roteco project. The aim of the project is the creation of a robotics teacher community where teachers can find the needed support in order to teach with robots in their classes.

However, an issue is also to motivate teachers to join and participate in online communities. According to Ling et al. [12], most members of communities remain passive and only 4–10% of members produce more than 50–80% of the messages and resources shared. The reasons for not participating are for example: lack of time, isolated work, lack of reflection on their practice, lack of technical support, pressure from state-mandated standards, pre-existing mistrust directed at the university and preferences for face-to-face interaction [11]. Other studies claim that the main reason to participate at online communities is to be part of a group and not for sharing information [8]. The cited studies are however not focused on educational robotics and STEAM. With the present document the aim is to focus on the reasons why teachers join an educational robotics community.

The present document describes therefore the Roteco project and the results of a questionnaire sent to the community members in spring 2020. The specific research questions addressed in the questionnaire are:

- Why do teachers join a community around educational robotics?
- How do they contribute to it?

2 Roteco Project

Roteco is a project promoted by three institutions (SUPSI, Wyss Institute of ETH Zurich/University of Zurich, and EPFL) and is funded by the Swiss Academies of Sciences and the ETH-Board.

2.1 The Goal of the Project: A Community of Teachers

The goal of the Roteco project is to create a vibrant community of school teachers in the field of robotics and computational thinking who share classroom practices around educational robotics, are informed about the new development in the field and offered trainings in the use of the robots. To support the start and building of a community of teachers, the Roteco project designed, developed and animated a web platform under the domain name Roteco.ch. There are several platforms for teachers and for educational robotics. Roteco.ch is unique since it is in the local Swiss languages (German, French and Italian), developed considering and starting from specific needs expressed by teachers with the possibility to give birth to a bottom-up development of teaching tools that could support the needs of the teachers. Besides the platform, Roteco also offers training courses for teachers, educational resources and participates at different events to promote educational robotics.

2.2 The *Roteco.ch* Platform

The Roteco project started in May 2018 with the design of a web platform for Swiss teachers. In January 2019 the platform was ready to be tested and by March 2019 the first Swiss teachers were invited to join the platform after a dedicated training program.

Roteco.ch is organized with a home page that describes the project and gives an overview as well as explanation of the platform. Most of the information is available without registration (e.g. news, events, courses, articles from educational robotic experts, resources, etc.). A login is only needed to access the learning activities and for the social interactions between teachers (names, emails, profile, comments, likes, etc.). By June 2020 *Roteco.ch* hosted more than 440 learning activities in the three platform languages, had more than 500 active teachers and more than 70 courses related to educational robotics and AI. More than 100 news items, 90 events and 45 stories complete the offer of contents for the teachers' community (Fig. 1). The content for *Roteco.ch* is developed in the three official languages of Switzerland: German, French and Italian. The three institutions that promote the Roteco project are also leading the community animation and management in the three languages. Five university of teacher education and three major associations dedicated to the promotion of sciences joined the project in order to enhance and expand the impact on the local communities of teachers in the different Swiss cantons. In 2020 the Roteco team decided to open the access to teachers and experts of educational robotics, AI and Computational Thinking from abroad in order to enrich the Swiss national community with open collaboration and discussion with worldwide teachers.

The Roteco.ch platform is designed and developed through a constant involvement of teachers and members of the community and some effort has been invested in creating an identity for teachers that they could proudly share. It is a flexible platform that adapts to the needs of teachers. The more the community uses it, the better it can be improved according to teachers' requests. Teachers can search for learning activities as well as create and share their own activities. They can search for other teachers filtering by language, expertise, subject taught, etc., and have access to news and best practices published by experts. Teachers also have access to free downloadable books and learning resources (e.g., cooperative adventures like an R2T2 Remote Collaborative Mission, escape games with robots, unplugged activities, etc.).

Fig. 1. Overview of the *Roteco.ch* contents

In the Roteco project there are different roles. The *Project Leader* is responsible to drive the entire project and is assisted by three *Community Managers* who animate the teachers and are organized by language (Italian, French and German). Another important role is the *Expert*. *Experts* contribute to *Roteco.ch* by providing best practices, training for the teachers, as well as share content such as books, learning activities ideas and pedagogical resources. The *Experts* can be found and easily contacted by teachers within the platform. The *Ambassadors* are leading teachers and "pioneers" who are able to invite other teachers to join the Roteco community and motivate them to go beyond their natural comfort zone as well as start to introduce robotics into their lessons. Their main role is of a "teacher/user". Each teacher is invited to share lesson plans, to comment and vote on any content that is shared on the *Roteco.ch* platform. They are invited to participate in events, challenges, cooperative adventures and trainings.

2.3 The Roteco Trainings

A continuing professional development training for teachers in educational robotics and computer science has also been set up. These training courses take place in three different regions of Switzerland (the cantons Vaud, Ticino and Zurich), in three different languages, and are made possible thanks to the collaboration of several Swiss continuing education institutions. They have been held every year since 2018. Their objective is to have the participating teachers experience educational robotics and its potential in order to reassure and accompany them. Indeed, teachers often tend to believe that educational robotics is too difficult and/or that they do not have sufficient skills to integrate it into their classrooms (low self-efficacy) [5].

During the three days of training, the *Thymio II* robot as well as different didactic activities are presented. The activities also focus on social content and transversal competences and not only on technical elements. The activities combine robotics with other STEAM disciplines and especially with transversal competences such as cooperation, communication and creativity. In addition to providing resources and content teachers can integrate in their classrooms, the Roteco training aims for a positive and enthusiastic change in the perceptions of teachers about robotics [13], so that they understand the benefits of integrating the new concepts [1].

On the last day of the training, a shared event for all three training sites is organized: an *R2T2 Remote Collaborative Mission.* The teachers are all in different geographical locations and have to cooperate, program and control 16 *Thymio II* robots, located at EPFL in Lausanne and connected remotely. In order to carry out the different phases of the mission, the teachers must not only program correctly, but also collaborate with the other teachers from the different language regions. These courses encourage the creation and growth of a true community of teachers with a common passion and specific interest in pedagogical innovation as well as the adoption of new practices related to educational robotics in the classroom.

2.4 The Roteco Resources

The Roteco platform contains several didactic resources from a variety of sources:

- Peer-created resources, to allow teachers to share their activities and co-create together. In this case, the factor of valuing and belonging to a community is emphasized. These activities can be "liked" and commented on by the community.
- Resources from universities of teacher education and other institutional partners active in the field of educational robotics.
- Resources specially created by the Roteco team for the needs of the community. Thus, depending on the seasons and circumstances, the Roteco team creates; for example, activities like the "Robotics Winter Olympic Games" during the winter or "Learning

at home" activities during the COVID-19 period. The Roteco team is always trying to adapt to the needs of the teachers.

These resources support different styles of teaching and are different in nature, for example, on the platform there are unplugged activities that do not require a PC or robotic equipment, activities with visual programming languages, and making-based activities meant to teach different subjects while using many diverse robots and tools.

2.5 Social Media, Newsletter and Participation at Events

In order to foster the creation of a community, Roteco also has a presence on Twitter, Facebook and YouTube. Furthermore, every two months a Roteco newsletter is sent to the community members to inform them about new courses, events and developments in the field of educational robotics. Roteco is not only present online but also participates at many events and conferences for teachers. This is to have an opportunity to meet teachers face-to-face, promote the use of educational robotics into schools and to learn about the new developments in the field in order to share these findings on the platform with the Roteco community.

3 Survey

3.1 Sample Description

In May 2020 a survey was conducted with all members of the Roteco community. In the sample there were 87 participating teachers (51 male, 36 female). The surveyed teachers belonged to three different language regions of Switzerland (20 German speaking, 31 French speaking and 36 Italian speaking). The majority of the respondents taught in primary school (26) or in secondary school (25). The others taught in pre-primary school (8), in a high school (7), in a university (7), in special education schools (5) or in a professional school (4). Additionally, there were participants who answered that they do not teach (5). The subject(s) taught by the respondents were varied: 50 taught technology and media related subjects, 43 mathematics, 27 natural sciences, 24 languages, 19 geography, 17 foreign languages, history or arts, 16 sport, and 8 music or religion. These numbers are explained as in Switzerland a teacher often teaches more than one subject, especially in primary school where teachers teach almost all of the disciplines. The majority of respondents were experienced teachers as most had been teaching between 11 to 20 years (28) or 20 plus years (23). In the sample, there also were participants new to the teaching profession who were enrolled in a university for teacher education studying to become a teacher (2) and a new teacher with less than one year of experience (1). The number of teachers who taught for 1 to 5 years (12) and 6 to 10 years (16). There were also "pioneers" and many teachers advanced with their digital skills as 55 indicated they were experts in the use of ICT, 21 felt intermediate and only 11 labeled themselves as beginners. Most teachers already participated in a training of educational robotics (67).

3.2 Procedure

Data were collected using a questionnaire developed in-house. All teachers registered on *Roteco.ch* received it in their mother tongue (German, French or Italian). The survey was structured around two questions: "Why do teachers join a community around educational robotics?" and "How do they contribute to it?" More specifically the questionnaire included questions about their use of *Roteco.ch* (e.g., how often do you use the activities on Roteco?), their behavior on *Roteco.ch* (e.g., how often do you interact with other teachers on Roteco?), their motivation to be part of a community (e.g. why did you join Roteco?) and questions about the usability of *Roteco.ch* (e.g. which technical problems did you have on Roteco?). The questionnaire also included their demographics, mother tongue, disciplines taught and level of expertise in educational robotics. It was sent out via newsletter and website. The format of the questions included open question items as well as 6-point Likert scale questions (e.g., 1 = never to 6 = often) and simple Yes/No question items. It had a length of 38 questions with an estimated time of response of 5–10 min. There was no pilot prior to the submission. A copy of the questionnaire can be downloaded at following link: https://www.roteco.ch/wp-content/uploads/2021/01/Questionnaires-Roteco.zip (in Italian, French or German).

3.3 Results

Almost all participants (72/87) answered that they joined Roteco in order to remain informed about robotics (Fig. 2). The other three main reasons were: to be inspired by colleagues (60), to obtain didactic materials (50), and to be part of a community (48). Lesser common reasons for joining Roteco were the ones that needed more effort from the participants namely to communicate with other teachers (43), share didactic materials (39), as well to find events (34) or courses (29). Some teachers (8) also answered that the reason they joined Roteco was that it was requested of them in their training course.

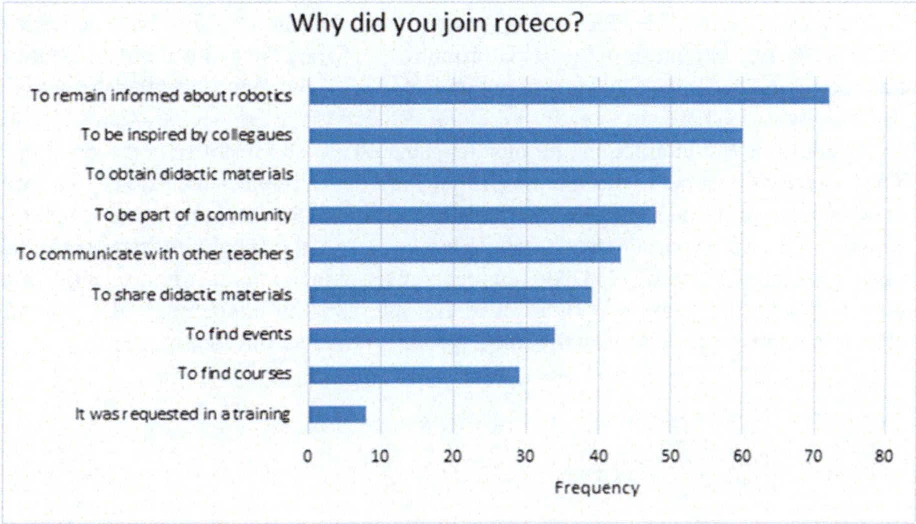

Fig. 2. Reasons to join *Roteco.ch*

Figure 3 shows that the participants in the survey do not use Roteco very often. When asked the reason why they do not use Roteco often, the most common answer was: "I do not have time" (39), followed by "I do not do robotics in my class" (7) or "I have just registered on *Roteco.ch*" (5). Some teachers also answered that "it is too difficult" (5), "there are no activities for the robots that I use" (4) and "that are too many other resources online" (3).

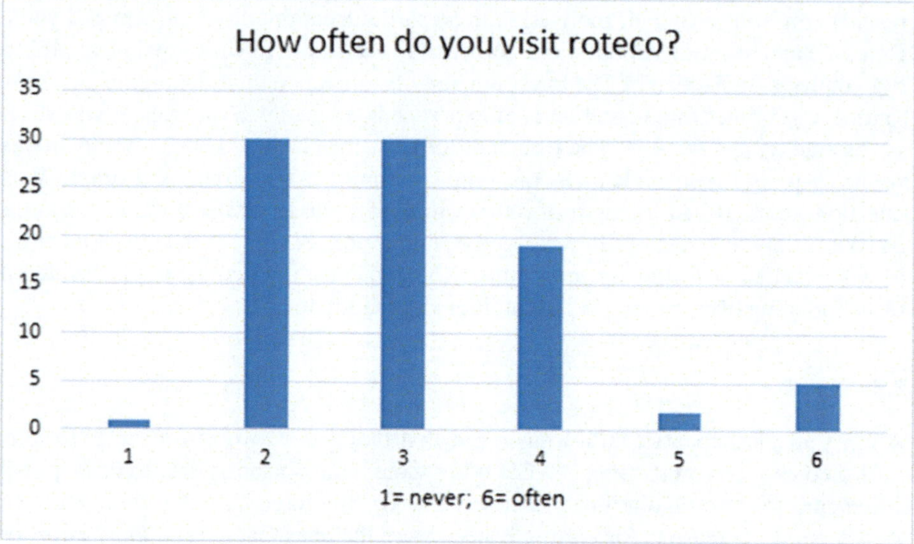

Fig. 3. How often do you visit Roteco?

A similar situation can be found when asked if they share activities on Roteco (Fig. 4). The majority in fact has never shared an activity on Roteco (51). The main reason given is the "lack of time" (33). Teachers also mentioned that they "do not have activities to share" (20), are "beginners in the field of robotics" (15), are "not sure about the quality of their activities" (15), have just registered on Roteco (7), and mentioned that the activities that they want to share are already present on Roteco (5).

In addition, the interaction amongst teachers on Roteco is not so common (Fig. 5). There were 51 teachers who answered that they never interact on Roteco. The main reasons were still "lack of time" (31), followed by "I do not know other teachers on Roteco" (23). Some participants also "did not know that they could interact" (11), "have just registered on Roteco" (11), "do not feel comfortable to interact online" (10), "do not do robotics in their classes" (9), are just "not interested in interacting" (4), "use other channels to interact" (1) or say that they "prefer to interact physically" (1).

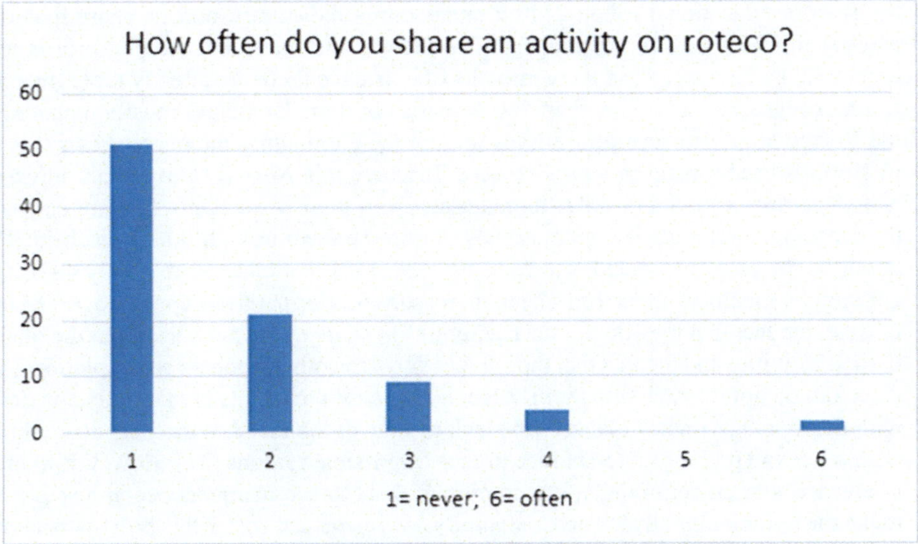

Fig. 4. How often do you share an activity on *Roteco.ch*?

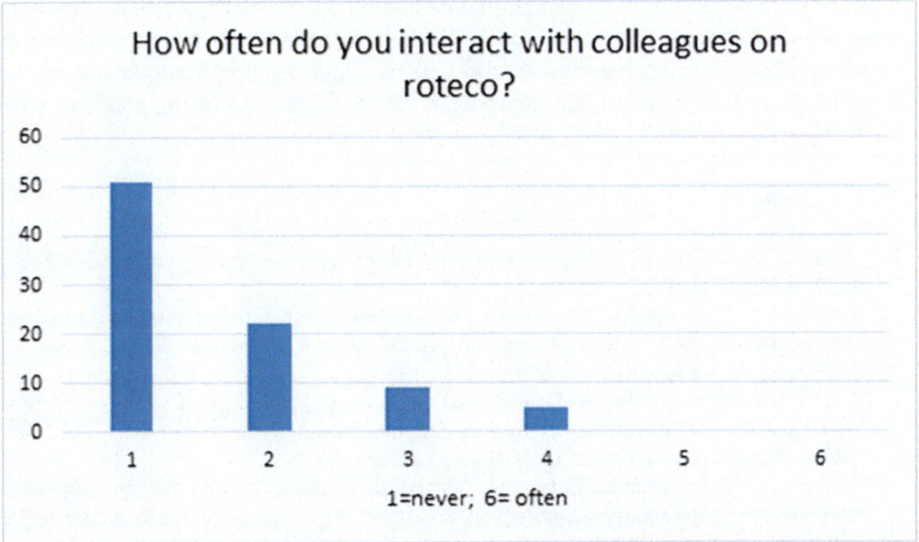

Fig. 5. How often do you interact with colleagues on *Roteco.ch*?

4 Conclusion

In order to increase the presence of educational robotics in schools and foster therefore the acquisition of the digital skills necessary for our society, the Roteco team developed a collaborative platform where teachers can remain updated with the new activities in the field of educational robotics, find open-source didactic resources, share materials and get in touch with other teachers. The first results show that the platform is used especially by "pioneers"; in a few months of existence there are already more than 500 teachers registered and more than 400 activities in three languages mostly prepared by the Roteco team. However, the survey among the community members shows that the platform is used mostly in a passive way. Teachers join *Roteco.ch* to remain informed and to get inspiration and didactic materials from their peers. Only few are active on the platform sharing their own materials or communicating with others teachers. This confirms the results founded also by Ling et al. [12] that claimed that only 4–10% of community members are active. The main reasons for this passivity appear to be: the lack of time, the fact that they do not have activities to share or are not sure about the quality of their activities and the fact that they do not know the other teachers on the platform and therefore do not interact with them. Also, in this case the results are partially consistent with the literature review that sees the lack of time or the fact that teachers do not know each other and prefer to interact face to face as possible reasons [11]. However, in order to create a vibrant community, it is important that Roteco members become active and make the community alive. The *Community Managers* can foster this process but some strategies to make teachers more active are needed. To react to this last point the Roteco team has thought to implement more social features on the Roteco platform in order to allow teachers to get to know each other more. For example, by giving the possibility to "follow" a teacher as well as organizing more live events and contests for the community. Knowing each other a little more could also help to have less barriers in sharing their own activities, since teachers know who will read and use their materials. In order to monitor the impact of these new features, statistics of the use of the platform will be collected and analyzed.

References

1. Alimisis, D.: Educational robotics: open questions and new challenges. Themes Sci. Technol. Educ. **6**, 63–71 (2013)
2. Atmatzidou, S., Demetriadis, S.: Advancing students' computational thinking skills through educational robotics: a study on age and gender relevant differences. Robot. Auton. Syst. **75**(Part B), 661–670 (2016)
3. Bocconi, S., Chioccariello, A., Dettori, G., Ferrari, A., Engelhardt, K.: Developing Computational Thinking in Compulsory Education – Implications for Policy and Practice. Publications Office of the European Union, Luxembourg (2016)
4. Bocconi, S., Chioccariello, A., Earp, J.: The Nordic approach to introducing computational thinking and programming in compulsory education. Report prepared for the Nordic@BETT 2018 Steering Group (2018)
5. Chevalier, M., Riedo, F., Mondada, F.: Pedagogical uses of Thymio II: how do teachers perceive educational robots in formal education? IEEE Robot. Automat. Mag. **23**, 16–23 (2016)
6. de Koster, S., Kuiper, E., Volman, M.: Concept-guided development of ICT use in "traditional" and "innovative" primary schools: what types of ICT use do schools develop?: concept-guided development of ICT use. J. Comput. Assist. Learn. **28**(5), 454–464 (2012)
7. Dionne, L., Lemyre, F., Savoie-Zajc, L.: Vers une définition englobante de la communauté d'apprentissage (CA) comme dispositif de développement professionnel. Revue des sciences de l'éducation **36**(1), 25–43 (2010)

8. Ellis, D., Oldridge, R., Vasconcelos, A.: Community and virtual community. Ann. Rev. Inf. Sci. Technol. **38**, 145–186 (2004)
9. Grossman, P., Wineburg, S., Woolworth, S.: Toward a theory of teacher community. Teacher College Record **103**(6), 942–1012 (2001)
10. Hedges, H.: Blurring the boundaries: connecting research, practice and professional learning. Cambridge J. Educ. **40**(3), 299–314 (2010)
11. Hur, J.W., Brush, T.A.: Teacher participation in online communities. J. Res. Technol. Educ. **41**(3), 279–303 (2009)
12. Ling, K., Beenen, G., Ludford, P., Wang, X., Chang, K., Li, X., et al.: Using social psychology to motivate contributions to online communities. J. Comput. Mediated Commun. **10**(4) (2005). https://urldefense.proofpoint.com/v2/url?u=https-3A__academic.oup.com_jcmc_article_10_4_JCMC10411_4614451&d=DwIDAw&c=vh6FgFnduejNhPPD0fl_yRaSfZy8CWbWnIf4XJhSqx8&r=RBo6AuZoqPUqdfIxMkY29PXcQu3yKI3zYibZM3SORus&m=9IdBESk6rKAYZGo1uNWr3hCzHoVVTYrpPWHT4zcd-ao&s=4dxQQRlSmfWkZXPk2N8qVhTOfoBRsrMZTBblJYTpTnc&e=
13. Negrini, L.: Teachers' attitudes towards educational robotics in compulsory school. Italian J. Educ. Technol. **28**(1), 77–90 (2020)
14. Negrini, L., Giang, C.: How do pupils perceive educational robotics as a tool to improve their 21st century skills? J. e-Learn. Knowl. Soc. **15**(2), 77–87 (2019)
15. Park, I.-W., Han, J.: Teachers' views on the use of robots and cloud services in education for sustainable development. Cluster Comput. **19**(2), 987–999 (2016)
16. Richardson, V., Placier, P.: Teacher change. In: Richardson, V. (ed.) Handbook of Research on Teaching, 4th edn., pp. 877–904. American Educational Research Association, Washington, DC (2001)
17. Vescio, V., Ross, D., Adams, A.: A review of research on the impact of professional learning communities on teaching practice and student learning. Teach. Teacher Educ. **24**, 80–91 (2008)

There is No Such Thing as a "Trial and Error Strategy"

Chiara Merisio[1,2], Gilda Bozzi[1,2], and Edoardo Datteri[1](✉)

[1] Department of Human Sciences for Education, RobotiCSS Lab (Laboratory of Robotics for the Cognitive and Social Sciences), University of Milano-Bicocca, Milan, Italy
c.merisio@campus.unimib.it, {gilda.bozzi, edoardo.datteri}@unimib.it

[2] Yunik, No-Profit Association, Milan, Italy

Abstract. The concept of "trial and error strategy" is pervasively used in the literature on robot programming: in particular, it is often claimed that some novices use a trial and error strategy, while others use a more rational and reasoned strategy. Here it is argued that the concept of "trial and error" is of little use for the analysis of the strategies adopted by novices and experts in programming robots. Indeed, in a certain sense, all problems and sub-problems faced by programmers are addressed in a trial and error *and* rational fashion. Moreover, that very concept does not play any meaningful role in revealing the richness of the criteria used by novices and experts to orient themselves in the vast space of the possible solutions to their problems. These claims will rely on an artificial-intelligence-inspired construal of the concepts of "problem" and "strategy" and will be substantiated by a preliminary analysis of the various sub-problems involved in robot programming.

Keywords: Problem-solving strategies · Psychology of programming · Programming problems · Heuristics

1 Introduction

Inquiring into the problem-solving strategies adopted by novices while they learn to program robots is important for several reasons, especially under the commonly held assumption that robot programming can stimulate problem-solving abilities [1–4]. These reasons concern evaluation and generalization of learning, and the design of educational activities. *Evaluating* whether learning has taken place may require one to assess changes in the nature or exercise of novices' programming strategies. *Designing* robot programming activities for novices may involve a reflection on what programming strategies teachers want them to develop or absorb. Studying robot programming strategies is also essential to stimulate *generalization* of what novices learn in a programming course, defined in terms of the ability to apply algorithm design strategies to other problem-solving contexts.

The study of *computer* programming strategies flourished in the second half of the XX century [5, 6] in connection with cognitive scientists' interest towards human

M. Malvezzi et al. (Eds.): EDUROBOTICS 2021, SCI 982, pp. 190–201, 2021.
https://doi.org/10.1007/978-3-030-77022-8_17

problem solving [7]. More recent works analyze *robotic* programming strategies [8, 9]. In some of these studies it is claimed that novice robot programmers may use a so-called "trial and error" (TE from now on) strategy. The TE strategy, in their opinion, is not the only programming strategy novices can use: other learners are claimed to use a more "reasoned", non-TE strategy. The distinction between the TE and the "reasoned" strategy may resonate in experts and teachers who observed novices learn to program. Some novices, when initially confronted with a novel programming task, quickly alternate between coding and testing phases. This may be classified as a TE process – indeed, they usually make a lot of errors. Other novices, rather differently, spend substantial time in reasoning about the "right" code before trying it. This may be classified as a "reasoned", non-TE process.

Contra these claims, it will be argued here that the distinction between a TE and a non-TE (reasoned, rational) strategy is too blurred and vague, and hence of little use for characterizing the problem-solving strategies used by novices while they program robots. More precisely, this paper has three goals.

- First, it will be argued that, in a general sense, all robotic programming is TE *and* reasoned, at the same time. There is no distinction between the two strategies, *contra* what is claimed by the authors cited above.
- Second, it will be argued that the TE/non-TE distinction – assuming that it holds, for the sake of the argument – falls short of capturing the nature of the distinction between the two categories of programmers identified by Gabriele et al. [9] and Gaudiello and Zibetti [8]. Programmers belonging to these two categories do not differ from each other in the adoption of a TE vs. a non-TE strategy, but in the cognitive strategies adopted in solving phases of the whole programming problem.
- Third, it will be argued that the TE/non-TE distinction fails to characterize the richness of the cognitive strategies used by novices to program robots.

These claims will be based on an AI-inspired definition of the concepts of "problem" and "strategy", and on the results of an exploratory empirical analysis of the micro-problems involved in collaborative programming activities.

2 The "Trial and Error Strategy"

It has been often claimed that, while experts use domain-specific strategies to solve problems, novices tend to adopt a "trial and error" approach [10, 11]. The concept of trial-and-error strategy was central in Thorndike's theory of animal intelligence [12]. According to Thorndike, when animals face a novel problem, they often try possible solutions and select the most satisfactory one, discarding unsuccessful actions. The association between problems and successful solutions is strengthened in time by repetition. In Thorndike's observations, animals often found solutions by chance.

The concept of "trial and error" is widely used in the literature on robot programming [1, 8, 9, 13–27]. Notably, some scholars argue that novice programmers may use two strategies, a TE strategy and a non-TE strategy, which is called "reasoned", "rational", or "problem-focused" in the literature.

Gabriele et al. [9], for example, investigated the strategies adopted by high school students in adjusting the evolutionary parameters of a neural-network-based robot. They "identified two main strategies: 1) Solution focused strategy: Based on modifications of the evolutionary parameters over and over again (trials and errors); 2) Problem focused strategy: Based on detailed analysis of the assigned task, taking into account the performance that the robot had to do" (p. 14). This reconstruction implies that students adopting the TE strategy did not carry out a detailed analysis of the assigned task and of the desired performance.

Gaudiello and Zibetti [8] propose that children may use three types of strategies (they use the term *control heuristics*, defined as "general rules or strategies that guide our actions in a problem-solving task, and that take into account contextual constraints as well as the final goal of the task", p. 16) in robot programming. They are: 1) "Task driven heuristics, in which sequences of actions are applied by children using a trial and error strategy, primal to achieve the solution for the task by following an implicit procedure"; 2) "Knowledge driven heuristics, in which single actions are applied by children through a reasoned strategy, mainly to seek explicit information about the rules of the task"; and 3) "Awareness driven heuristics, by which actions are applied by children through both trial and error and reasoned strategies, mainly to establish the limits of the task and to assess their own understanding of the task" (p. 17).

Here the authors clearly counterpose a TE strategy (first category) to a "reasoned strategy" (second category), the third category somehow combining the first two. These categories are exemplified with reference to a robot control task which involved no programming: children had to manipulate some knobs so that a robotic arm lifted and put a sugar lump into a coffee cup. Some children "manipulated the knobs one after another, observing the corresponding motor activation and adjusting the movement in real time" (p. 20). This corresponds to the first category. Other children "manipulated each of the five knobs and observed the corresponding motor activation. They only started to solve the task once they had understood the one-to-one correspondence between each knob and the related motor movement" (p. 20). This is the reasoned strategy, which is explicitly claimed to be "not trial-and-error" (p. 20). According to the authors, it is "reasoned" because children tried to understand the function of the knobs before starting to use them to solve the task. Still other children "manipulated the knobs … observing the corresponding movement … in order to assess what they could and could not do with the robot. They only started to solve the task or proposed different or further tasks after this" (p. 20). These children "were more interested in becoming aware of what they could do and could not do, in terms of possibilities … and limitations" (p. 20).

Over and above the details of this taxonomy, it is worth pointing out that Gaudiello and Zibetti counterpose a TE strategy with a reasoned strategy, implying that the TE strategy is not reasoned (and vice-versa). A similar claim is made by Alimisis [27], who claims some novices go "beyond trial-and-error strategies and try out more rational approaches to problem solving that will be closer to the scientific methods" (p. 283).

Is it true that learners may use a TE or a non-TE (reasoned, rational, problem-focused) strategy when they try to solve robot programming problems? To address this question, a clarification on the concepts of "strategy" and "problem" is needed.

3 Strategies and Programming Problems: A View from Artificial Intelligence

People adopt strategies to solve problems. To understand what strategies are in a general sense, and what robot programming strategies may be adopted, it is essential first to understand what a (robot programming) *problem* is. According to Robertson [28], a problem "is a difference between where you are now (e.g., your vacuum cleaner has stopped sucking) and where you want to be (e.g., you want a clean floor). In each case 'where you want to be' is an imagined state that you would like to be in" (p. 1). Not all differences between current and desired states define a problem: as pointed out by Reese [29], they become problems when there is an obstacle in between. If no obstacle hinders progress toward a goal, "attaining the goal is no problem" (p. 200). In these circumstances, the agent "has the task of devising some action which may mediate between the existing and the desired situations" (p. 1) [30].

What counts as a *strategy* to solve problems? Quite surprisingly, even though the term "strategy" is so frequently used in the literature on learning, no clear definition is offered. Research on the conceptual foundations of cognitive science and Artificial Intelligence (AI), stemming from the pioneering works carried out by Allen Newell and Herbert Simon on human problem solving [7], offers resources to formulate more precise definitions of the concept of strategy. The AI conception of a problem-solving strategy relies on quite a technical characterization of what a problem is, which will be introduced here (avoiding unnecessary technicalities). For details, see [31]. Problems are classically defined in AI in terms of a starting state, a desired state, and a set of possible operations whose execution may change the current state. States are defined in term of certain parameters taking values, and describe possible situations at some level of abstraction. The state space is the set of all the possible states. Solutions are sequences of operations which "connect" the starting to the desired state. Identifying a solution requires the agent to explore the state space, and this may be done by searching for paths in the so-called decision tree. The decision tree for a problem starts from the initial state, which is the root node of the tree. Every node has branches corresponding to all the operations that can be carried out on it. Branches end on other nodes, which represent the states – called *successor* states – that are reached from the predecessor state after carrying out the corresponding operation. There may be leaf nodes, having no successors, and the same state can be represented by multiple nodes in the tree. Exploring such a decision tree involves starting from the root node and, via some search method (see below), finding the route which takes to the node representing the desired state.

What counts as a problem-solving *strategy* in this theoretical framework? As recalled above, solutions to AI problems are found using *search methods*, which may be blind or heuristic [31]. Blind search methods – such as the so-called breadth-first, or depth-first strategies – explore the state space in a "brute force" fashion: they consider all the possible paths until a "right" one is found. They may require significant time and memory resources. Heuristics are criteria that can be used to exclude parts of the state space from the search, or prioritize certain paths over others. The term "strategy", in AI, can be used to denote a heuristic search method for a particular problem.

Since no precise definition of the concepts of "strategy" and "problem" can be found in the robot programming literature, the AI-inspired definitions will be adopted in the ensuing analysis. What may count as a problem and a strategy in robot programming, then?

Intuitively, robot programming involves facing many problems defined at many levels of analysis. At the most general level, programmers face a problem that may be informally stated as: starting from the empty program, find the program that produces a given behavior. This *whole-program problem* is defined in terms of an initial state, which is represented by the empty program; a desired state, which is the program which produces the desired behavior; and a set of operations, which includes programming actions (e.g., inserting, deleting, modifying instructions) as well as physical actions involved in moving the robot, connecting it to the Internet, testing the program, etc. The state space is the set of all the possible programs (see Fig. 1).

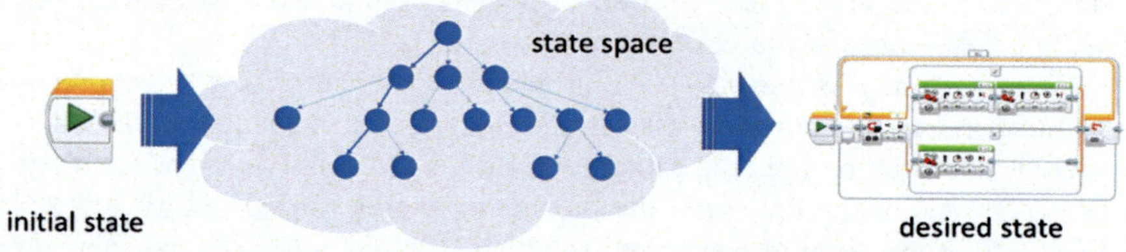

Fig. 1. The whole-program problem. The initial state is the empty program. The desired state is the "right" program. To formulate it, programmers must explore the state space including all the possible programs (represented as dots in a decision tree).

The whole-program problem is not the only problem faced by programmers. Indeed, it may be decomposed into many other lower-level problems which programmers may have to face in order to produce the "right" program. Purely notional examples of subproblems include the following.

- *Revising a program which does not work*. The initial state is the current program; the final state is the "right" program; the operations include coding and physical actions on the environment and on the robot.
- *Setting-up the testing environment*. The initial state is the current environment; the final state is the "right" environment; operators include physical actions on the environment and on the robot.
- *Specifying the desired behavior*. This may be a problem, in some cases. The initial state is a vague and underspecified mental representation of the desired behavior; the final state is a precise and fully specified mental representation of the desired behavior; cognitive operations transform the initial to the final representation.

To sum up. Programming a robot requires an agent to address a variety of problems. At a macro-level of analysis, they address the whole-program problem. But they also go through several programming phases which correspond to lower-level problems, being definable in terms of relations between initial states, desired states, and possible (physical or cognitive) operations.

What counts as a *strategy* in robot programming? In the AI-inspired framework adopted here, it is a heuristic search method which can be used to explore the state space defined by a given robot programming problem. Each problem involved in robot programming will likely have a set of potentially useful strategies. Different programmers may choose different strategies to address the same robot programming problem.

Given these analyses of the concepts of problem and strategy, let us return to the question introduced at the end of Sect. 2: is it true that novices may use a TE or a non-TE (reasoned, rational, problem-focused) strategy when they try to solve robot programming problems?

4 All Programming is Reasoned Trial and Error

If strategies are heuristic search methods, then the term "TE strategy" denotes a trial-and-error heuristic search method. Unfortunately, the research literature offers no clue about how a TE heuristic search method may work – i.e., about what criteria it may use to prune or prioritize paths in the decision tree. So, we are left with an intuitive understanding of this strategy: a TE strategy is a method to explore decision trees which involves considering possible solutions (i.e., possible paths in the tree), checking if they really are solutions and, in case they do not (which will count as an error), considering another possible solution. This intuitive conceptualization, however, implies that all programming strategies are TE. All heuristic search methods involve selecting paths in the tree, checking if they are solutions and, if they are not, selecting another path. This is how *all* search methods work. More generally, all problems are solved, by novices and by experts, by trying solutions and checking for errors (by actually running the program or, as experts typically do, by mental simulation). This is true even when it turns out that the agent made no error. Therefore, or the TE strategy is defined in more stringent and restrictive terms, or one must conclude that all strategies are TE strategies – which makes any talk of a TE strategy relatively useless – and, consequently, *contra* the authors cited before, that all novices, if they use a strategy, use a TE strategy to program.

Moreover, unless a more precise definition of TE strategy is given, one may legitimately argue that no distinction can exist between a TE strategy and a "reasoned" or "rational" strategy. Indeed, if a strategy is a *method* for exploring the decision tree, then it must provide some criterion to choose which path to explore. Be it based on "brute force" or on domain-specific branch-ranking functions, a criterion gives reasons to decide what solution to try next. Let us develop this idea by focusing on the whole-program problem, which is the problem of formulating the "right" program starting from nothing, and exploring the space of the possible programs.

If all novices use a TE strategy to program, as argued above, then they all use a TE strategy to address the whole-program problem. One may object that this claim incautiously obscures the reasonable distinction made by [8, 9]. As pointed out in Sect. 2, these scholars partition the class of novices into two (or three) categories. Some learners, according to these authors, enter the fray by quickly alternating coding and testing phases, thus making a lot of errors, while other learners allot substantial time in coding their solutions before testing them, thus likely making fewer errors (even though this needs not be the case, and grey zones are possible). The distinction between TE and non-TE strategy was supposed to capture this distinction. If all novices use a TE strategy, how can this distinction be saved?

To this question, we respond as follows. Let us concede that novices can be partitioned in the two behavioral groups identified by the authors cited above. Still, it is false that novices in the first class use a TE strategy, and novices in the second class use a rational, non-TE strategy. Novices in the second class try solutions and make errors, even though they spend more time than the others in the first phase. They use a rational *and TE* strategy. At the same time, it is hard to believe that novices in the first class try their solutions by chance. They spend less time in producing solutions, but this does not imply that this phase is carried out irrationally. They use a TE *and rational* strategy. Thus, novices in both classes use a *TE and reasoned* strategy, *contra* what is claimed by [8, 9].

Where, then, do the two classes of novices differ from one another? Arguably, in the way they deal with different phases of the robot programming process – using our AI-inspired conceptual framework, in the strategies used to address different programming sub-problems, a claim that we will try to substantiate in the following section.

5 Programming Stages and Strategies

5.1 Problems and Sub-problems in Robot Programming

Addressing whole-program problems requires one to deal with several sub-problems. This idea was explored in the last decades of the XX century by Ruven Brooks who, in the spirit of Newell and Simon's analyses of problem decomposition [7], and Polya's [32] four-stage reconstruction of problem-solving processes, proposed that programmers go through three broad distinct phases named understanding, method-finding, and coding [5]. Brooks' analysis is specifically related to computer programming. What sub-problems are involved in the programming of a *robot*? And what credit can be given to the idea that the strategies adopted by novices, while they deal with these problems, can be TE or not?

To address the first question, we carried out an exploratory study in 2019, which involved observation and video-analysis of some programming activities organized by the "EXPLORA" Museum in Rome (Italy). The activities involved a total of 36 children aged 9–10, organized in two sequential sessions. In each session, the children were organized into small groups (3–5 children per group). Each group had to solve the following problem. Some cards, each showing one of the letters of the word "EXPLORA", were spread on the floor in a non-linear order. Starting from letter "E", each group had to program a Lego MINDSTORM EV3 robot so that it passed through each letter of the word, in the correct order. The whole activity was video- and audio-recorded using four handycams. A screen capture software (FlashBack express) was used to record the coding activity of each group, and the facial expressions of the children were recorded using the computer webcams. Later, the various video sources were synchronized, so that the coding behavior could be easily related to the external behavior of the robot. The video structure was then annotated by one researcher (using the ELAN software [33], Max Planck Institute for Psycholinguistics, https://archive.mpi.nl/tla/elan) in a way loosely inspired to the so-called grounded theory [34]. The researcher, first, identified programming problems faced by all the groups (the marker for the occurrence of a problem was the detection of a difficulty, signalled by the verbal and non-verbal behavior of the pupils, according to the informal definition of problem provided in Sect. 3). Then, she organized them into a two-level hierarchy of problems loosely inspired by Brooks' [5] and Polya's [32] analyses of (programming) problem-solving processes. Table 1 shows the results.

Table 1. Taxonomy of problems and sub-problems

Problems	Sub-problems
Specifying the desired behaviour of the robot	- Understanding what behaviour the robot should display - Understanding explicitly stated constraints on the desired behaviour of the robot (e.g., places through which it should not pass, which are explicitly mentioned in the problem specification) - Making explicit constraints which are not mentioned in the problem specification (e.g., that the robot should move at a normal pace)
Designing the algorithm	- Decomposing the algorithm to be designed into sub-algorithms - Designing the algorithm
Coding the program	- Managing the interface and the programming area - Translation of the algorithm into a code - Writing the code
Experimenting on the program	- Setting-up the experimental environment - Launching the experiment - Observing and describing the behavior of the robot - Understanding the causes of the observed behaviour
Debugging the program	- Identifying the discrepancies between the desired and actual behaviour of the robot - Identifying the (software, hardware, environmental) causes of the discrepancies between the desired and actual behaviour of the robot

For the sake of simplicity, the taxonomy proposed here omits problems specifically devoted to the management of the group in the various phases of programming. The taxonomy is not linear, meaning that it is not meant to represent the temporal or logical sequence of programming problems faced by novices. Note also that the analysis proposed here is preliminary and explorative. From a methodological point of view, it could be improved along several lines. For this reason, the taxonomy shown in Table 1 is neither claimed to be complete nor correct, and will be revised in future studies. However, it suffices to substantiate the claim that programming involves addressing many sub-problems, in addition to the whole-program problem, and gives a first, preliminary clue about their nature. Let us bring this taxonomy to bear on the distinction between a TE and a non-TE strategy.

5.2 Sub-problems and Strategies

Each of the sub-problems shown in Table 1 can be defined in terms of a starting state, a desired state, and a set of operators. In some cases, these elements pertain to the domain of cognition. *Designing the algorithm*, for example, is the problem of building a mental representation of the "right" algorithm starting from nothing, and using cognitive processing operations. In other cases, the problem is defined at the level of digital representations. In *Writing the code*, for example, the initial state is the empty program (represented in some way in the computer's memory), the final state is a digital representation of the "right" program, and the operators are coding actions (e.g., moving a programming block from the palette to the program). In still other cases, the problem is physically defined (e.g., as in *Setting-up the experimental environment*). We propose here the thesis that all these sub-problems generate a decision tree, and that solving these problems requires one to find the "right" path using some heuristic search method, i.e., a strategy.

If this thesis is true, what credit can be given to the claim that novices may use a TE or a non-TE strategy to solve these sub-problems? As argued before, the very idea of a distinction between a TE and a "rational" or "reasoned" strategy is weak at best. Strategies provide criteria to explore decision trees: if a problem is solved through a strategy, it is not solved by chance. *Designing the algorithm* involves manipulating mental representations of an algorithm. Sure enough, this phase can be carried out more or less rapidly: the programmer may produce an algorithm in one second, or after ten minutes. In a *Writing the code* phase, the programmer may write a program in a one-shot fashion, showing a high level of confidence, immediately switching to one of the *Experimenting on the program* phases; or, they may produce a code more slowly and after a lot of corrections. However, both categories of programmers adopt a *TE and reasoned* strategy. They both adopt a TE strategy: they reach a possible solution (an algorithm, a code) after discarding potential alternatives in the state space, e.g., potential errors. And, they both adopt a *reasoned* strategy: they use some criterion to select that possible solution, a criterion which may determine more or less processing, thus being more or less time-consuming.

If it is true that all programmers use a *TE and reasoned* strategy to address the whole-program (Sect. 4) and the sub-problems involved in robot programming, then the distinction between a TE and a non-TE strategy vanishes. This makes the very concept of a TE strategy (as opposed to a non-TE one) inconsistent – hence the provocative title of this paper – and of no use to classify the problem-solving strategies involved in robot programming. *A taxonomy of programming strategies should indeed reveal the various criteria which are used by novice and expert programmers to explore the decision tree generated by each possible programming problem and sub-problem.* The TE/non-TE distinction is simply a non-starter for addressing this problem – TE and non-TE are not different criteria to explore decision trees.

Imagine, for example, two programmers differing from one another in the time needed to address a *Design an algorithm problem*, i.e., to produce a mental representation of a potentially "right" algorithm. This difference could be behaviorally signalled by the time spent in "thinking" before clicking on the palette of commands: the quick programmer immediately starts to code; the slow programmer lingers for a while.

The first programmer will solve the whole-program problem in a more hectic way, possibly making a higher number of errors, while the second one will seem to be more "rational". But, as argued here, this is not the right way to conceptualize the difference between these two programming styles. Both programmers are adopting a TE and reasoned strategy to solve the whole-program problem, for the reasons discussed in Sect. 4. And, both programmers supposedly used a heuristic search method – a TE and reasoned strategy, for the reasons discussed here – to address that instance of the *Design an algorithm problem* category. The interesting question, here, is what criteria the two programmers used to produce an algorithm among all the algorithms included in the state space. Possibly, the first programmer selected the first algorithm memorized in an ordered repertoire of algorithms built during previous programming experiences, while the second programmer adopted some additional criterion to exclude a certain number of algorithms in their repertoire. This is a purely notional hypothesis, which is made here only to exemplify the thesis that the TE/non-TE distinction, in addition to being inconsistent for the reasons discussed here, is a non-starter for classifying the richness of the heuristic search methods actually used in the many sub-problems which novices must address while they program robots.

6 Concluding Remarks

This paper has offered a conceptual and theoretical reflection on the concept of "trial and error strategy", which is pervasively used in the robot programming literature. As such, it was meant to be a purely theoretical contribution. In the spirit of Turkle and Papert's epistemological pluralism [35], we take it for granted that novices may use different programming styles, some of them being hectic and making a lot of errors, others "thinking" a lot before trying solutions. But it is wrong to say that novices in the first category adopt a "trial and error strategy" as opposed to a reasoned, rational one. All robotic programming is trial and error and reasoned at the same time, this making the very concept of "trial and error strategy" inconsistent – hence the title of this paper. This result may be taken to suggest that teachers and educators should avoid use of this concept in the analysis of the programming strategies adopted by novices, in favor of a more informative and technically precise taxonomy of the criteria used to explore the decision trees associated to the problems they face. Investigating the nature and content of this taxonomy, and suggesting ways to use it in the practice of teaching and education, are questions for future studies.

References

1. Barak, M., Zadok, Y.: Robotics projects and learning concepts in science, technology and problem solving. Int. J. Technol. Des. Educ. **19**(3), 289–307 (2009)
2. Benitti, F.B.V.: Exploring the educational potential of robotics in schools: a systematic review. Comput. Educ. **58**(3), 978–988 (2012)
3. Bers, M.U., Flannery, L., Kazakoff, E.R., Sullivan, A.: Computational thinking and tinkering: exploration of an early childhood robotics curriculum. Comput. Educ. **72**, 145–157 (2014)

4. Atmatzidou, S., Demetriadis, S., Nika, P.: How does the degree of guidance support students' metacognitive and problem solving skills in educational robotics? J. Sci. Educ. Technol. **27**(1), 70–85 (2018)
5. Brooks, R.: Towards a theory of the comprehension of computer programs. Int. J. Man-Mach. Stud. **18**, 543–554 (1977)
6. Brooks, R.: Towards a theory of the cognitive processes in computer programming. Int. J. Man-Mach. Stud. **9**, 737–751 (1983)
7. Newell, A., Simon, H.A.: Human problem solving. Prentice-Hall, New Jersey (1972)
8. Gaudiello, I., Zibetti, E.: Using control heuristics as a means to explore the educational potential of robotics kits. Themes Sci. Technol. Educ. **6**(1), 15–28 (2013)
9. Gabriele, L., Marocco, D., Bertacchini, F., Pantano, P., Bilotta, E.: An educational robotics lab to investigate cognitive strategies and to foster learning in an arts and humanities course degree. Int. J. Online Eng. **13**(4), 7–19 (2018)
10. Mayer, R.E.: Thinking, problem solving, cognition, 2nd edn. Freeman, New York (1992)
11. Hennessy, S., McCormick, R.: The General Problem Solving Process in Technology Education. In: Banks, F. (eds.) Teaching Technology, pp. 94–108. Routledge, London (1994)
12. Thorndike, E.L.: Animal Intelligence: Experimental Studies. The Macmillan company, New York (1911)
13. Hussain, S., Lindh, J., Shukur, G.: The effect of LEGO training on pupils' school performance in mathematics, problem solving ability and attitude: Swedish data. Educ. Technol. Soc. **9**(3), 182–194 (2006)
14. Highfield, K., Mulligan, J., Hedberg, J.: Early mathematics learning through exploration with programmable toys. In: Proceedings of the Joint Meeting of PME 32 and PME-NA XXX, vol. 3, pp. 169–176 (2008)
15. Nugent, G., Barker, B., Grandgenett, N., Adamchuk, V.I.: Impact of robotics and geospatial technology interventions on youth STEM learning and attitudes. J. Res. Technol. Educ. **42**(4), 391–408 (2010)
16. Alimisis, D.: Robotics in education & education in robotics: shifting focus from technology to pedagogy. In: Proceedings of the 3rd International Conference on Robotics in Education, pp. 7–14 (2012)
17. Eguchi, A.: Educational robotics to promote 21st century skills and technological understanding among underprivileged undergraduate students. In: 5th IEEE Integrated STEM Conference, pp. 76–82 (2015)
18. Elkin, M., Sullivan, A., Bers, M.U.: Programming with the KIBO robotics kit in preschool classroom. Comput. Schools **33**(3), 169–186 (2016)
19. Gomoll, A., Hmelo-Silver, C.E., Šabanović, S., Francisco, M.: Dragons, ladybugs, and softballs: girls' STEM engagement with human-centered robotics. J. Sci. Educ. Technol. **25**(6), 899–914 (2016)
20. Kaya, B.E., Newley, A., Deniz, H., Yesilyurt, E., Newley, P., Standards, G.S.: Introducing engineering design to a science teaching methods course through educational robotics and exploring changes in views of preservice elementary teachers. J. College Sci. Teaching **47**, 66–75 (2017)
21. Eguchi, A., Uribe, L.: Robotics to promote STEM learning: educational robotics unit for 4th grade science. In: ISEC 2017 - Proceedings of the 7th IEEE Integrated STEM Education Conference, pp. 186–194 (2017)
22. Ioannou, A., Makridou, E.: Exploring the potentials of educational robotics in the development of computational thinking: a summary of current research and practical proposal for future work. Educ. Inf. Technol. **23**(6), 2531–2544 (2018)
23. Jung, S.E., Won, E.: Systematic review of research trends in robotics education for young children. Sustainability **10**, 905 (2018)

24. Kim, C., Yuan, J., Vasconcelos, L., Shin, M., Hill, R.B.: Debugging during block-based programming. Instruct. Sci. **46**, 767–787 (2018)
25. Scaradozzi, D., Screpanti, L., Cesaretti, L.: Towards a Definition of Educational Robotics: A Classification of Tools, Experiences and Assessments, Smart Learning with Educational Robotics, pp. 63–92. Springer, Cham (2019)
26. Anwar, S., Bascou, N.A., Menekse, M., Kargar, A.: A systematic review of studies on educational robotics. J. Pre-College Eng. Educ. Res. **9**(2), 19–42 (2019)
27. Alimisis, D.: Teacher training in educational robotics: the ROBOESL project paradigm. Technol. Knowl. Learn. **24**(2), 279–290 (2019)
28. Robertson, S.I.: Problem Solving. Perspectives from Cognition and Neuroscience, Second Edition. Routledge, London and New York (2017)
29. Reese, H.W.: Cognitive and behavioral approaches to problem solving. In: Hayes, S.C., Hayes, L.J., Sato, M., Ono, K. (eds.), Behavior analysis of Language and Cognition. Context Press (1994).
30. Duncker, K.: On problem-solving. Psychol. Monographs **58**(5), 1–113 (1945)
31. Russel, S., Norvig, P.: Artificial Intelligence: A Modern Approach, 4th edn. Pearson, London (2020)
32. Polya, G.: How to Solve it, 2nd edn. Princeton University Press, Princeton (1957)
33. Andersson, R., Sandgren, O.: ELAN analysis companion (EAC) for time-course analysis of ELAN-annotated. Data J. Eye Move. Res. **9**(3), 1–8 (2016)
34. Glaser, B.G., Strauss, A.L.: The Discovery of Grounded Theory: Strategies for Qualitative Research. Aldine, Chicago (2018)
35. Turkle, S., Papert, S.: Epistemological pluralism: styles and voices within the computer culture. Signs J. Women Cult. Soc. **16**(1), 128–157 (1990)

Educational Robotics: School to University, Examples of Interconnected Longitudinal Individual Student Routes

Anton Yudin[1,2(✉)], Andrey Vlasov[1], Dmitriy Zatekin[1,2], and Stepan Lapshinov[2]

[1] Bauman Moscow State Technical University, Moscow, Russia
skycluster@gmail.com, vlasovai@bmstu.ru, zatekindv@student.bmstu.ru
[2] Technical Education Center, SBVEI "Vorobyovy Gory", Moscow, Russia
stepan.lapshinov@yandex.ru
http://www.bearobot.org

Abstract. This paper presents several examples of students' (12–17 y.o.) results of successful long-term iterative educational process in digital fabrication laboratory environment united by the idea of building original robotic solutions to annually changing competition tasks. Presented examples demonstrate that within educational robotics domain finding long-term uniting ideas to bring education to the next level of implementation is possible and relatively easy. Being a complex multidisciplinary subject, educational robotics has big potential to play one of the important roles in the ongoing transformations in school.

Keywords: Educational robotics · STEM · Technological literacy · Digital fabrication · Engineering · Supplementary education · Longitudinal study · Customized learning

1 Introduction

Educational robotics (ER) is a complex subject, a field with a lot of opportunities and as many challenges.

Lammer et al. in [1] states that despite numerous efforts of researchers and engineers, teachers and educators face the problem of "navigation" in approaches, frameworks and tools available to them in the ER.

Some challenges and problems of the ER are common for general teaching, especially sciences and STEM, and some are quite specific for the field.

Similarities with common general education include constant decrease of students' interest throughout the world [2–4]. According to [5], in developed countries like Germany physics teaching in secondary school does not inspire or motivate, foster creative thinking or produce applicable skills. Worse, it fails to provide a framework for understanding the basis of modern technology and often discourages students from pursuing scientific or engineering studies.

As for specific challenges, although it is believed that the ER started as a discipline more than 50 years ago [6] with the work of Papert on programming

M. Malvezzi et al. (Eds.): EDUROBOTICS 2021, SCI 982, pp. 202–215, 2021.
https://doi.org/10.1007/978-3-030-77022-8_18

language to control a robot, today it is still on its way to maturity. Research and its outcomes are still criticized for being limited in terms of scale [7], being too narrow in terms of inclusion of wider range of students with different interests [8], etc. It is also evident for authors of [9] that technology has become more important than learning activity in today's classrooms, as well as teachers concentrate more on already interested young learners failing to attract others.

Yet one of the key challenges expressed in [10] concerns technology and its potential to transform education which has often been stated, though for various reasons, this potential is yet to be harnessed as expected. The particular domain of research on educational technology clearly demonstrates that what researchers state as key research questions are quite often unrelated to teachers' needs.

The criticism is justified as most of the research papers try to concentrate on isolated aspects of teaching and learning in a limited time-frame, simplifying the rest being out of scope. The simplification often leads to further narrowing of students' activity options by using predefined solutions to technical challenges. Construction kits of different sorts using blocks are indeed easy to start in STEM, but they tend to last and generally result in breaking of creativity cultivation or as addressed in [11] creativity preservation as all the kids have this capacity but most loose it with education.

The complex problem the authors of this paper try to address: how to gradually teach technical sciences, the engineering way of thinking without limiting creativity of students. This way each study evolution step means discovery of a more in-depth layer of precision possible to be introduced to the already known technical aspects of the produced artifacts and solutions. Such approaches are rather rare to be successful and thus non-mainstream due to fundamental complications of assisting students in their hands-on project-oriented work for an applied result, rather than providing the well-known structured solutions by the teacher. Examples of such work in educational robotics domain: [12–17]. A general example of a good design result according to the described approach would be a programming environment in the form of Squeak [18] and Pharo [19] systems, both based on Smalltalk language advocated by Alan Kay.

The latest reports show that school systems did not succeed on the large scale in adequate reaction to the COVID-19 pandemic enforced period of distant education. Such circumstances result in urgent accelerated new didactic emerging without traditional academic discourse [20]. Experts believe that at these times of uncertainty best practices will form foundations of the Fourth Industrial Revolution educational paradigm via self-reflection processes within a limited time frame of several years.

With introduction of protective measures against COVID-19 spread in schools many younger students were forced to spend extreme amounts of time behind a computer. A large study of 56 876 schoolchildren (K-12) and 160 895 parents by the Russian Institute of Developmental Physiology [21] during the period of March to May 2020 shows that very common approach was to transfer the offline material to online form without changes, which resulted in extreme stress and fatigue for children, teachers and parents.

The COVID-19 pandemic in fact intensifies the rather popular "blocks-based" approaches to be widely used and thus leads to the threat of further adoption of "limited by design" methods in general education without caution of breaking creativity on the large scale.

Besides other consequences, now it becomes evident for the authors that hands-on practice in the described circumstances of isolation does not work well for the ER in the authors' context of supplementary education for schoolchildren, even for the most active ages 10–14 and for the older ones (14–18). While remarkable efforts are applied to organize virtual laboratories to simulate the real environment, the results show very low student activity of the named age groups. This renders blended approaches to be far more likely to succeed for ER rather than distant ones.

Global consensus is that school curricula needs to be changed for the new reality brought by the Fourth Industrial Revolution. Lifelong learning came in existence and brings flexible learning trajectories, which do not follow the traditional fixed time periods and knowledge-based assessment. Completion rates are thus substituted by measuring demonstrated proficiency in competency-based models.

One of the basic and important competences for the new economy is technological literacy which could be taught in schools within the ER [22] in connection to exact sciences. For this introduction to be successful the ER has to respect interests of all children and provide long-term hands-on programs integrated with sciences [9].

This paper adds to the series of previously published self-reflecting summarizing work on the ER by the authors [23,24] with several examples of students' results of a successful long-term educational process in digital fabrication laboratory environment united by the idea of building original robotic solutions to annually changing competition tasks. This is done in hope of motivating further research on adoption of the discussed hands-on approach to school curricula of the near future.

2 Individual Route Cases

The examples include 2 individual routes, taken and mapped by 2 students in the context of supplementary (after-school) education. Both routes relate to robotic competition [25], digital fabrication environment and started at their age of 12 for both students. Evolution of the content is systematic from simple to complex and covers 5 years of studies.

Most of the first steps are taken by novices at the age of 10–12, this is also true for these examples. This is the stage of mastering technologies and the first trials. The students had no prior experience in engineering and technology.

Methodological considerations and description of the educational environment were presented in earlier works by the authors [14–17,23,24,26] and are same for this research. Elements of the approach are presented in video form [30].

2.1 Robot Route

Figure 1 presents all of the steps taken by the first student. On the right side one can see robotic solutions to competition's tasks. Robots are distinguished by being remotely controlled or autonomous, and by levels of their complexity. Some inner stages of development are presented for the latest step on the left side of the figure. To achieve better results in competition participants are forced to improve their strategy which means iterative development of ideas realized in several robotic modifications for the same edition. Novices can not do that before realizing such importance in annual projects' implementation and participation in competitions.

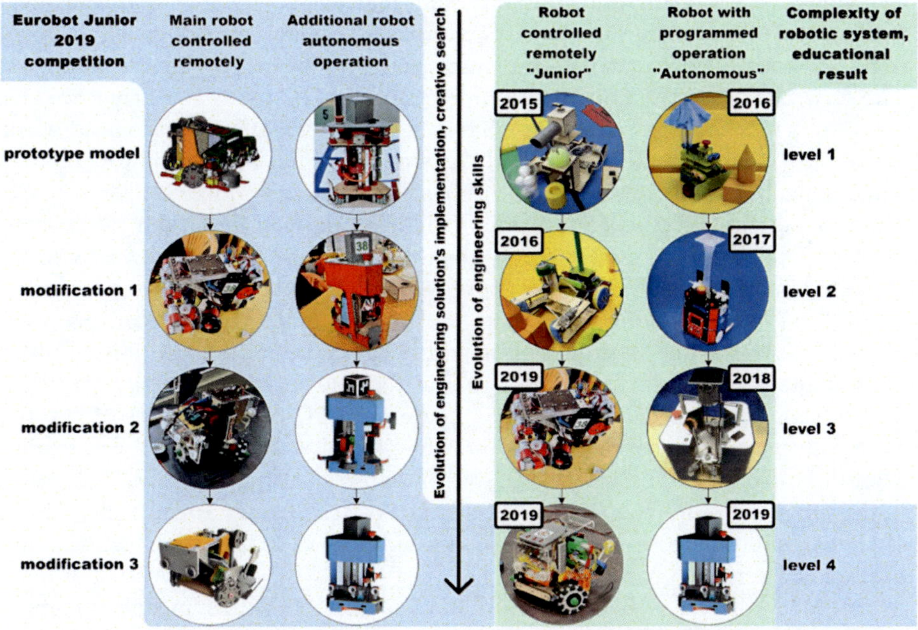

Fig. 1. Evolution of educational results for longitudinal project-oriented studies

At the time of the route's start, planning is not of special importance for students. Rather, we are talking about implementation of ideas, more often non-working or working unreliably. At this stage, it is important to maintain the spirit and motivation and complete the studies by presenting a robot in competitions among similarly novice teams.

One of the authors began this work in 2015 with participation in the Eurobot Junior competition. At that time, without the use of any CAD system, the "main" robot was designed to be controlled from console. When implemented it could move around the playing field and perform primitive tasks, pushing game elements with its body and thereby moving them to the desired zone and position. The "additional" robot was made on the basis of a Lego Mindstorms controller with modernized mechanics – an original solution assembled by hand. Its task was limited to one specific action that required autonomous navigation and programming. Robots' parts were made with partial use of a laser cutting machine and later modified with hand tools. The drawings of the parts for robots were primitive and were developed in the laser cutting machine's control program.

In 2016, participation continued in the junior league, but there was a qualitative change in understanding of the actions necessary to prepare for competition and organize work. After gaining competitive experience, efforts were made to improve the reliability of solutions used, both in mechanics and in programming. As before, the bodies and elements of the mechanisms of robots were manufactured on a laser cutting machine, but combinations of different materials began to be used more often. 3D-printed parts appeared in the design. The "main" remotely controlled robot received improved motion control driver electronics. The control scheme used a transitional version for DC motors: a driver was used for robot's movement, which allowed precise change for the wheels' speed of rotation, while other actuators still used the simplest "digital" circuit (on/off). The "additional" autonomous robot received new ultrasonic range sensors to determine its position relative to the side of the playing field, but their use was very limited. The robot was controlled by the 8-bit AtMega32u4 microcontroller. The digital Dynamixel XL-320 servomotors were used for the robot's movement.

In 2017, a single robot was designed and manufactured according to the restrictions of the "additional" and could participate in both the junior and the "autonomous" leagues. The robot was fully modeled in CAD SolidWorks, and a significant number of its parts was made on a 3D-printer. Parts of the body were also made on a laser cutting machine, and its chassis' base was made on a milling machine. This was the first experience of using such a CNC machine in design. Control for the robot utilized the AtMega328 microcontroller (as part of the Arduino platform). Ultrasonic rangefinders were used for navigation and obstacle detection.

In 2018, the team had two autonomous robots, which meant a complete transition to the autonomous league. Robots applied technology to design and manufacture very similar to the previous year's robot. But now, both robots also used self-fabricated original PCBs, a step was taken to improve electronics solutions. The robots were modeled in the SolidWorks CAD. Production of parts from plywood and plexiglass for the robots was carried out using a laser cutting machine; elements of complex design and fasteners were made using a 3D-printer. The printed circuit boards were designed in the Eagle CAD and then manufactured on a precise milling machine.

In 2019, it was decided to unite experiences and skills of students from both routes and to return to the junior league and show both the high level of technical performance and the corresponding result in competition. Two robots were developed: remotely controlled and autonomous. The mechanics and electronics of these robots were superior to those of the previous years. The robots complemented each other well, went through 3 cycles of redesign and refinement, which ultimately led to high efficiency in performing all possible tasks and to an original technical solution that was not repeated among other competitors. The robots were developed in the SolidWorks CAD, their design features have changed several times during development. The PCBs were designed in the Eagle CAD and Altium Designer. The required parts were manufactured on a laser cutting machine, 3D-printer, milling and turning (lathe) machines.

Table 1 summarizes evolution of the design processes and study results.

Table 1. Annual comparison of robotic projects

	2015	2016	2017	2018	2019
Number of robots: remotely controlled/autonomous	1/1	1/1	0/1	0/2	1/1
Fabrication machines used in development	laser	Laser, 3D-printer	Laser, milling, 3D-printer	Laser, milling, lathe, 3D-printer	Laser, milling, lathe, 3D-printer
Computer aided design programs used in development	CNC software, CorelDraw	CorelDraw	CorelDraw SolidWorks	Eagle, SolidWorks	Eagle, Altium Designer, SolidWorks
Types of motors used for movement	collector DC motor	Servo motor, collector DC motor	Stepper	Stepper	Stepper, collector DC motor
Drive type	Differential	Differential	Differential	Differential	Omni-directional
Control type	Electro-mechanical, Lego NXT	Electro-mechanical, Arduino	Arduino	Arduino	Arduino
Strategy sophistication (1 - low, 3 - high)	1	2	2	2	3
Overall effectiveness (1 - low, 3 - high)	1	2	2	1	3

Figures 2, 3, 4 demonstrate several technical aspects of the autonomous robot developed for 2020 edition of the competition's rules.

The Figs. 2, 3, 4 reflect all of the expertise gained in robot design after 5 years of hands-on project-oriented studies in digital fabrication laboratory environment previously described in [24]. It is the 6th step-year of improvement.

While early steps of the route tend to be more "intuitive" and show the lack of thinking about the winning strategy and iterative design processes, the later steps show how previously gained competition experience leads to understanding of importance of planning before doing.

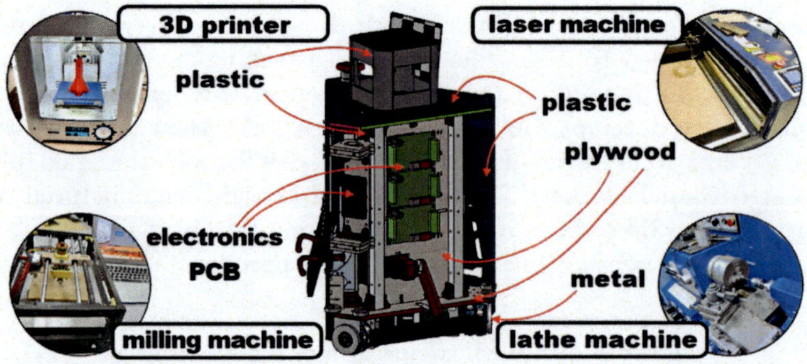

Fig. 2. Fabrication machines and technologies used in the robot's construction

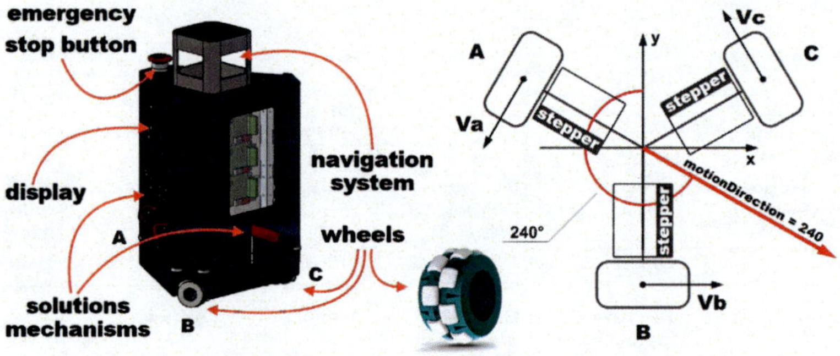

Fig. 3. The robot's design overview and its movement principle

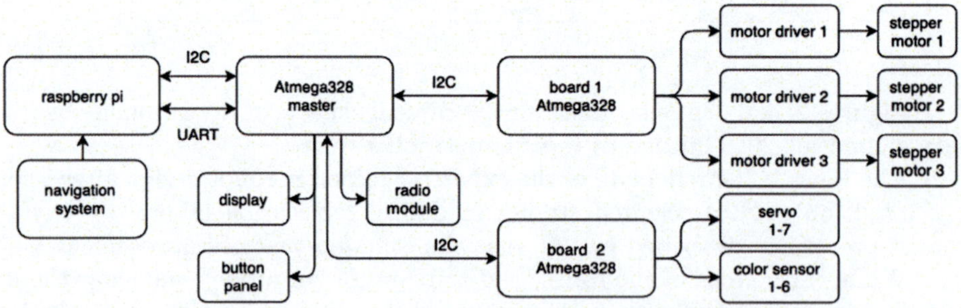

Fig. 4. Example of the robot's control composition

2.2 Ultrasonic Navigation Route

The learning route consisted of 5 steps, each taken annually:

1. (2015) An educational workbench for practical study of ultrasonic (US) waves. Several students were asked to study physics of processes that underlie widely available cheap US rangefinders. The project was meant to help physics lessons at school, providing other schoolchildren with a tool to experiment with US properties. The developer team (see Fig. 5) was composed of 4 participants (2 of which mapped the routes presented in this paper).
2. (2016) A system for monitoring fill level of garbage containers. Practical application of the US. As a result of the joint work of schoolchildren, university students and business, an operational prototype was created based on the project rendered in the learning process [26].
3. (2017) A prototype positioning system for mobile robots reproducible in educational digital fabrication laboratories using standardized purchased devices (ultrasonic rangefinders HC-SR04 and Arduino). The need to create a positioning system naturally arises among participants in mobile robots' competitions (for example, Eurobot [25]). Figure 6 (number 1) partly presents the development process for electronics.
4. (2018) Preparation of scientific work for admission to university on the basis of the previously developed prototype. This step extended theoretical side of the project, documentation and improved stability of operation. The first version of the positioning system was made ready for competition utilization. During the development process, some ready-made modules were replaced with self-made ones, and some additional were added. The resulting system significantly surpassed the prototype. The explanatory note was highly regarded by academic community and helped the author to enter university. Figure 6 (number 2, 3) partly presents the design of electronics, Arduino was substituted with a 32-bit ARM microcontroller. Figure 7 gives the idea of the modeling tasks carried out as well as the principle of operation for the system.
5. (2019) Another iteration of development of an original positioning system for mobile and underwater robots from scratch with self-developed parts. The complex nature of the problem behind the design allowed the author to continue the development, now in the status of university student. Knowledge gained at university made it possible to improve the algorithms for the positioning system, and along with the help of modern CAD systems (MATLAB, Altium Designer, SolidWorks), to achieve the stable operation of a self-made solution at a new level.

Figures 7, 8, 9 all show current state of development for the positioning system in question and its possible usage in the competition.

While early steps of the route tend to be more "intuitive", current state of understanding, based on the gained experience, leads to more modeling (Fig. 9). Which in turn adds practical use and real-life application to theoretical academic knowledge provided in university lectures.

Fig. 5. Initial project with research of ultrasonic properties, demonstration of results

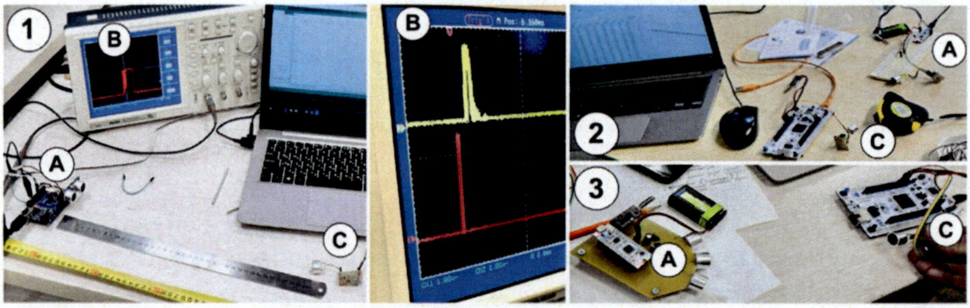

Fig. 6. Development and debugging of receivers and transmitters, design of electronics

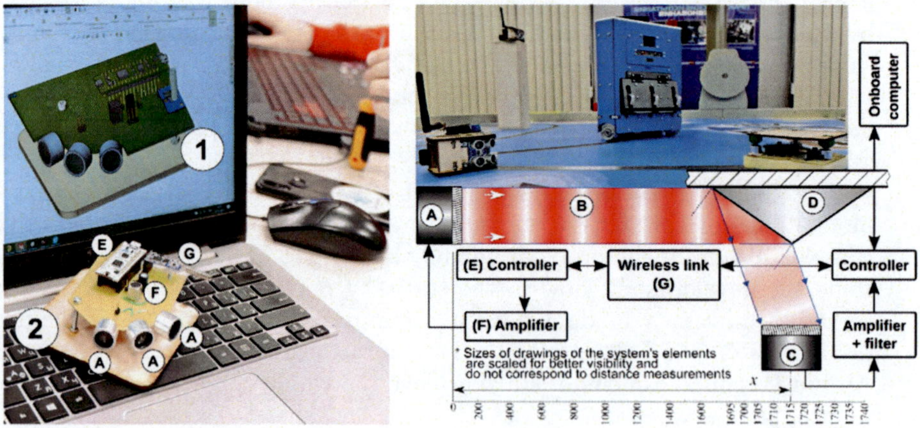

Fig. 7. Modeling of positioning system's elements and implementation

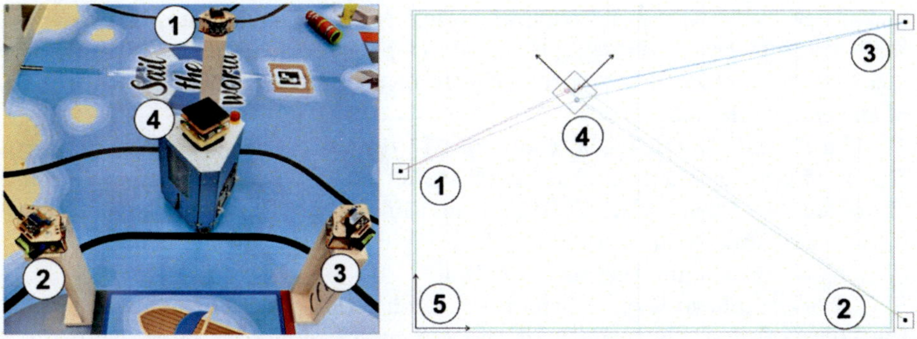

Fig. 8. Overview of the navigational system, its components and environment

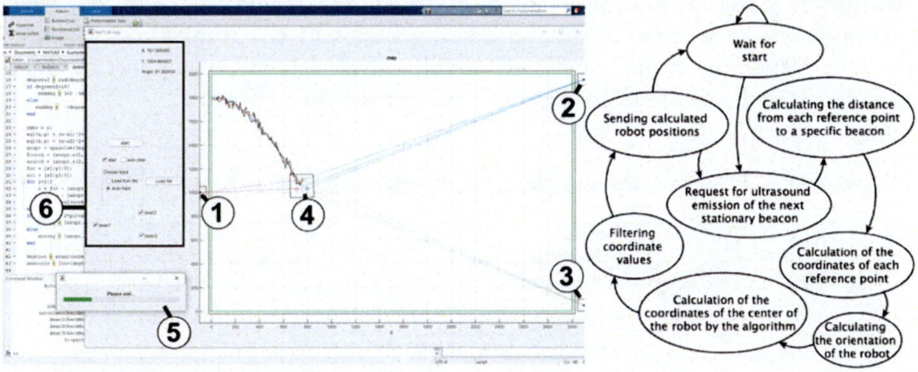

Fig. 9. Computational modeling of automated operation, movement and positioning

3 Conclusion

The described example material reveals possible educational routes based on the iterative process of hands-on practice and gradual addition of theoretical explanation for the sake of systems' precision, increase of the design's effectiveness. This material can serve as proof for the underlying principles and the idea of teaching organization in the context of digital fabrication laboratory at school, college or university. It could be used for further development of a new approach in the framework of blended learning for novice students starting to practice engineering at school and continuing their studies further in the same manner naturally improving their skills and increasing knowledge with the help of the ER.

The content of the "robot" route is naturally renewed each year with introduction of new competition rules, yet good practices from previous years help students to build further and dive deeper in understanding theories of operation for better results. Outcomes and built devices, even for teams of students united for the competition sake, could be documented for additional assessment of gained knowledge (substituting or complimenting common examination), ex. [27–29]. Skills could be assessed during the system's presentation and/or in competition, based on the achieved solution's effectiveness, implementation and complexity of the chosen challenge.

The second, "navigation" route clearly reveals capacity of the ER (competition robotics in particular) in approaching the teaching of common scientific knowledge and principles. Started from simple physics within several annual iterations it evolved in a useful engineering application which could be used to demonstrate academic theories in action.

Table 2 emphasizes one of the many similarities between implementations of the technical systems in both of the presented longitude routes (data in the table taken from the first route, from later steps on motion of the omni-wheeled drive). During design process students usually would face dilemmas leading them to several possible solutions. Unlike an ideal "mathematical" solution, each of them would have advantages and disadvantages at the same time. Resolution of such situations would mean invaluable advancement in engineering type of mentality and general thinking self-organization for a student.

Table 2. Specific properties of different approaches to motion control of a mobile robot

Atmega328 (32 KB)	Calculating on the go	Pre-calculation
Used microcontroller memory	5%	80%
Move command execution speed	50 ms	10 ms
Direction accuracy of movement	0.01°	1°

Indeed any part of a typical robotic system could become the root of a lasting educational process. Many topics in physics, math and programming could be taught while building a robot or similar complex technical device in a manner "try yourself" - "learn" - "do better". The results of such educational process could also extend other sciences with tools, developed by students themselves and thus naturally engaging them outside of the engineering field.

Such approach could gradually form connection between school and university as it is based on practice and could be extended with theory of the needed complexity. This means that transition from school to university, when both accept and use this approach, could become less stressful for students.

Being a complex multidisciplinary subject, the ER has big potential to play one of the important roles in the ongoing transformations in school. With presented examples it must be clear that, within the ER, finding long-term uniting ideas to bring education to the next level of implementation is possible and relatively easy.

Development of approaches that imply exercise of creative skills in the process of engineering educational work of varying degrees of complexity presents a large field for finding new opportunities to improve the effectiveness of education.

Acknowledgements. Some results were obtained in the framework of the State task #0705-2020-0041"Component's digital transformation methods' fundamental research for micro- and nanosystems".

References

1. Lammer, L., Vincze, M., Kandlhofer, M., Steinbauer, G.: The educational robotics landscape exploring common ground and contact points. In: Merdan, M., Lepuschitz, W., Koppensteiner, G., Balogh, R. (eds.) Advances in Intelligent Systems and Computing, vol. 457, pp. 105–111. Springer, Cham (2017). https://doi.org/10.1007/978-3-319-42975-5_10
2. Johnson, W.C., Jones, R.C.: Declining interest in engineering studies at a time of increased business need. accreditation board for engineering and technology, pp. 243-252 (2006)
3. Morales, L.-M., Maillet, K.: E-learning consequences of the declining interest in engineering studies in Europe. In: World Conference on E-Learning in Corporate, Government, Healthcare, and Higher Education (2009)
4. Kim, S.: New research shows declining interest in STEM (2018). https://www.govtech.com/education/k-12/New-Research-Shows-Declining-Interest-in-STEM.html. Accessed 30 Oct 2020
5. Education and Skills 2.0: New Targets and Innovative Approaches, WEF (2014). http://www3.weforum.org/docs/GAC/2014/WEF_GAC_EducationSkills_TargetsInnovativeApproaches_Book_2014.pdf. Accessed 30 Oct 2020
6. Altin, H., Pedaste, M.: Learning approaches to applying robotics in science education. J. Baltic Sci. Educ. **12**(3), 365–377 (2013)
7. Bredenfeld, A., Hofmann, A., Steinbauer, G.: Robotics in education initiatives in Europe - status, shortcomings and open questions. In: Proceedings of SIMPAR 2010 Workshops, pp. 568-574 (2010). ISBN: 978-3-00-032863-3
8. Alimisis, D.: Educational robotics: open questions and new challenges. Themes Sci. Technol. Educ. **6**(1), 63–71 (2013)
9. Lammer, L., Lepuschitz, W., Kynigos, C., Giuliano, A., Girvan, C.: ER4STEM educational robotics for science, technology, engineering and mathematics. In: Merdan, M., Lepuschitz, W., Koppensteiner, G., Balogh, R. (eds) Robotics in Education. Advances in Intelligent Systems and Computing, vol. 457, pp. 95–101 (2017). https://doi.org/10.1007/978-3-319-42975-5_9
10. Artificial Intelligence in Education: Challenges and Opportunities for Sustainable Development. United Nations Educational, Scientific and Cultural Organization. (2019). https://unesdoc.unesco.org/ark:/48223/pf0000366994. Accessed 30 Oct 2020
11. Robinson, K.: Changing Education Paradigms (2010). https://youtu.be/zDZFcDGpL4U. Accessed 31 Jan 2021
12. Jäggle, G., Vincze, M., et al.: iBridge - Participative Cross-Generational Approach with Educational Robotics. In: Lepuschitz, W., Merdan, M., Koppensteiner, G., Balogh, R., Obdržálek, D. (eds) Robotics in Education. Advances in Intelligent Systems and Computing, vol. 829, pp. 263–274 (2019). https://doi.org/10.1007/978-3-319-97085-1_26

13. Jäggle, G., Merdan, M., Koppensteiner, G., Brein, C., Wallisch, B., Marakovits, P., Brunn, M., Lepuschitz, W., Vincze, M.: Project-based learning focused on cross-generational challenges. In: Merdan, M., Lepuschitz, W., Koppensteiner, G., Balogh, R., Obdržálek, D. (eds.) Robotics in Education. Advances in Intelligent Systems and Computing, vol. 1023 (2020). https://doi.org/10.1007/978-3-030-26945-6_14
14. Yudin, A., Sukhotskiy, D., Salmina, M.: Practical mechatronics: training for mobile robot competition. In: 6th International Conference on Robotics in Education, RiE 2015, pp. 94–99 (2016). ISBN: 978-2-9700629-5-0
15. Yudin, A., Salmina, M., Sukhotskiy, V., Dessimoz, J.-D.: Mechatronics practice in education step by step, workshop on mobile robotics. In: 47th International Symposium on Robotics, ISR 2016, pp. 590–597 (2016)
16. Yudin, A., Vozhdaev, A., Sukhotskiy, D., Salmina, M., Sukhotskaya, T., Sukhotskiy, V.: Intensive robotics education approach in the form of a summer camp. In: Alimisis, D., Moro, M., Menegatti, E. (eds.) Educational Robotics in the Makers Era. Advances in Intelligent Systems and Computing, vol. 560. Springer, Cham, pp. 246–250 (2017)
17. Yudin, A., Vlasov, A., et al.: Challenging intensive project-based education: short-term class on mobile robotics with mechatronic elements. In: Lepuschitz, W., Merdan, M., Koppensteiner, G., Balogh, R., Obdržálek, D. (eds.) Robotics in Education. Advances in Intelligent Systems and Computing, vol. 829, pp. 79–84 (2019). https://doi.org/10.1007/978-3-319-97085-1_8
18. Open-source Smalltalk programming system. https://squeak.org. Accessed 30 Oct 2020
19. A pure object-oriented programming language and a powerful environment. https://pharo.org. Accessed 30 Oct 2020
20. Adamsky, A.: Key institutions and characteristics of the new school, keynote speech (2020). https://www.facebook.com/instituteeureka/videos/1218651705148628/. Accessed 30 Oct 2020. (in Russian language, speaker's presentation at 5:33
21. Bezrukikh, M.: Distance learning experience. Analysis of the organization of the educational process. Research results. Institute of Developmental Physiology, Russian Academy of Education (2020). https://drive.google.com/file/d/14Cj2-805yzEzjxMZH-H4MOv5mDlIgkQj/view. Accessed 30 Oct 2020. (in Russian language)
22. Jäggle, G., Lammer, L., Hieber, H., Vincze, M.: Technological literacy through outreach with educational robotics. In: Kampis, G., Karsai, I., Szathmáry, E. (eds.) Advances in Artificial Life. Darwin Meets von Neumann. Advances in Intelligent Systems and Computing, vol. 1023, pp. 114–125 (2020). https://doi.org/10.1007/978-3-030-26945-6_11
23. Yudin, A., Vlasov, A., Salmina, M., Shalashova, M.: Evolution of educational robotics in supplementary education of children. In: Merdan, M., Lepuschitz, W., Koppensteiner, G., Balogh, R., Obdržálek D. (eds.) Robotics in Education. Advances in Intelligent Systems and Computing, vol. 1023, pp. 336–343 (2020). https://doi.org/10.1007/978-3-030-26945-6_30
24. Yudin, A., Vlasov, A. Perspectives of Educational Robotics in the Ongoing Technological Transformations (to be published)
25. Eurobot: International Students Robotic Contest. https://www.eurobot.org. Accessed 09 Mar 2020

26. Yudin, A., Kolesnikov, M., Vlasov, A., Salmina, M.: Project oriented approach in educational robotics: from robotic competition to practical appliance. In: Merdan, M., Lepuschitz, W., Koppensteiner, G., Balogh, R. (eds.) Advances in Intelligent Systems and Computing, vol. 457, pp. 83–94. Springer, Cham (2017)
27. Yudin, A., Semyonov, M.: Distributed control system for a mobile robot: tasks and software architecture. In: Communications in Computer and Information Science, vol. 161, pp. 321–334. Springer, Heidelberg (2011)
28. Demidov, A., Kuturov, A., Yudin, A., Krasnobryzhiy, B., Chistyakov, M., Borovik, R.: Autonomous mobile robot development in a team, summarizing our approaches. In: Obdržálek, D., Gottscheber, A. (eds.) Communications in Computer and Information Science, vol. 156, pp. 168–179. Springer, Heidelberg (2011)
29. Kuturov, A., Yudin, A., Pashinskiy, I., Chistyakov, M.: Team development of an autonomous mobile robot: approaches and results. In: Obdržálek, D., Gottscheber, A. (eds.) Communications in Computer and Information Science, vol. 161, pp. 187–201. Springer, Heidelberg (2011)
30. Educational Robotics Augmented Channel. https://www.youtube.com/channel/UCURM98qJYQhx8saQs8gx4lA. Accessed 01 Jan 2021

Exploring the Use of Educational Robotics in Primary School and Its Possible Place in the Curricula

Effransia Tzagkaraki, Stamatios Papadakis(✉), and Michail Kalogiannakis

Department of Preschool Education, University of Crete, Rethymno, Greece
{stpapadakis,mkalogian}@uoc.gr

Abstract. This paper is a brief review of the literature on the use of educational robotics in primary school. The purpose is to explore the application of robotics and, more specifically, the advantages robotics offers to students, the challenges that arise from its application, and what is its place in the curricula. Educational robotics is an innovative and useful tool. It positively affects critical thinking, computational thinking, problem-solving, algorithmic thinking, creativity, and collaboration. The literature reveals that difficulties arise either at the technical level or due to teachers' lack of relevant knowledge or the lack of relevant provisions for their effective integration into primary school curricula.

Keywords: Educational robotics · Primary/Elementary school · Curriculum

1 Introduction

Traditional teaching and learning practices need to be transformed to respond to the modern workplace's evolution and prepare the future labor force [1, 2]. The teacher also needs to transform his role and, consequently, his profession and act as a researcher, as new knowledge is continually being revised [3]. However, none of them would be essential if these changes are not focused on the curriculum [4].

Using and teaching with robotics has been increasingly studied in recent years, as research data have shown useful recommendations and positive results for students [4–8] and teachers [9]. Positive research data exist even for preschool children [10–12].

In general, educational robotics has been linked to essential skills, cognitive and social, such as metacognition, computational thinking, critical thinking, problem – solving, creativity, collaboration, and making decisions. Furthermore, educational robotics are harmonized with constructivism, constructionism, and social constructivism, so they could be a basis for applying new teaching methods compatible with the demands of the 21st century.

M. Malvezzi et al. (Eds.): EDUROBOTICS 2021, SCI 982, pp. 216–229, 2021.
https://doi.org/10.1007/978-3-030-77022-8_19

2 Method

2.1 Research Strategy

This paper is a review of the literature on the use of educational robotics in primary education. The aim is to draw and record conclusions through this research to clarify aspects of the application and educational robotics contribution. Readers thus could make judgments about the necessary theoretical and empirical positions.

Based on the aim, the following research questions arise:

1. What are the outcomes that arise from the application of educational robotics activities for students?
2. What challenges are faced by using educational robotics in practice?
3. What is the possible place of robotics in curriculum design?

2.2 Research Design

The value of a literature review lies in the presentation and interpretation of data from individual surveys. While there may be doubts about the researcher's subjectivity, a systematic collection of research data is always useful [13–17].

According to the study's purpose, we started by shaping the research questions, focusing on primary school educational robotics. We searched for instructions for conducting a review, and then the search criteria were determined for the relevant studies. The databases or other sources were identified, and we searched using keywords. In the 'Results' chapter, we summarize the findings from the various studies, and we interpret and discuss the results taking into account the limitations of our study, proposing topics for further investigation [13, 15–17].

The search for inclusion or exclusion of a study that constituted the eligible population was based mainly on the title, abstract, mesh terms, language, and publication date. Mesh terms used in search engines (Google Scholar, Research Gate, Science Direct) were 'educational robotics,' 'primary/elementary education,' 'curriculum.'

Of the 52 papers found in total, seven were excluded after reading the title and abstract, and 24 were excluded after reading the full text. In the end, 21 papers were selected as eligible (Fig. 1).

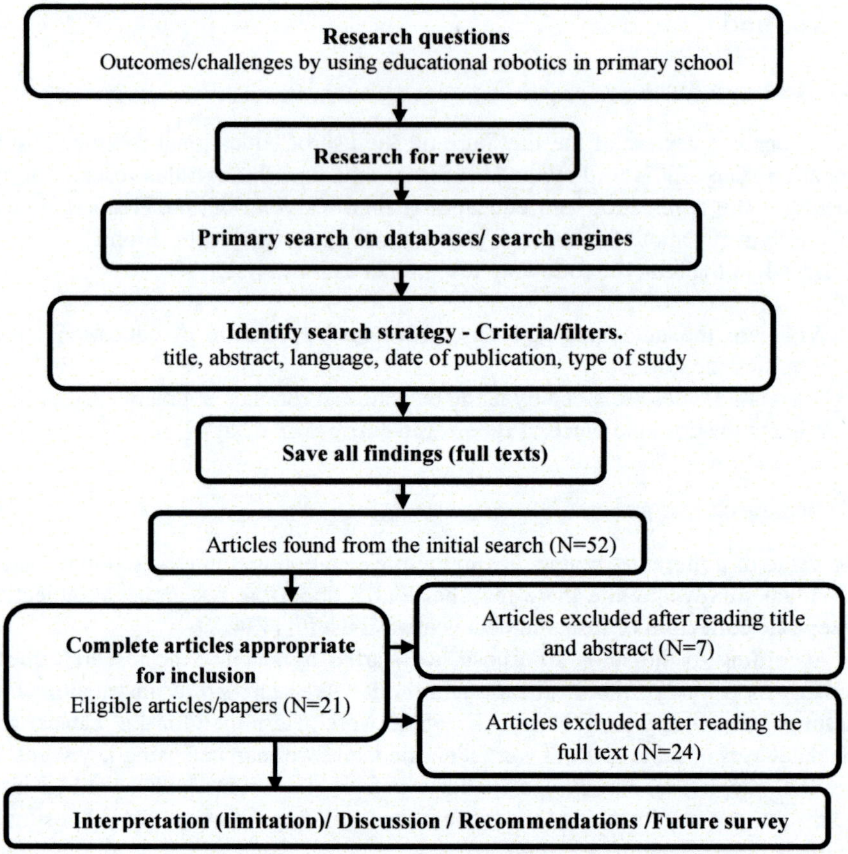

Fig. 1. Flow chart for literature review [15, 16]

3 Results

3.1 What Outcomes Arise for Students by Applying Educational Robotics Activities

Robotics is an innovative and attractive tool for students of all ages. It is an exciting field for children as they satisfy their curiosity about the world [4, 5, 18]. According to research, compared to the traditional teaching method, robotic activities actively engage students and encourage autonomous learning and motivation [7, 19, 20].

With robots, students can represent not concrete concepts by improving their understanding of school content [1] but also the whole world by making conceptual models [19] and growing 'visual and conceptual understanding' of an idea [18]. Besides, the possibility of using robotics in subjects such as history, arts, and literature can help students improve their knowledge and skills and make their teaching more exciting and engaging through active interaction [4, 9].

Learning is enhanced when a research process is followed [19]. Based on a project, students design, program, and interact with the robot. They need to think about the problem, evaluate and re-evaluate their ideas at each stage of the process. To find the solution, students observe, cope with the external conditions, and identify the effects on the robot's behavior, try and correct mistakes [7, 21]. In this context, kids foster their critical thinking and the most important, adopt the ability to apply their knowledge in a wide range of similar situations.

Students improve their creativity and imagination through the process of programming a robot. From the beginning, they plan what they will build based on the goal they want to accomplish. They make assumptions and imagine their operation. At the same time, they can do self-criticism and receive feedback on their effort [22].

As students actively engage in robotics projects, following a series of logical steps combine fun with learning, which adopts a metacognitive and humanistic character. As they are active and experimenting, shape their understanding of the world. Knowledge becomes more valuable and lasts longer in memory than whether they are just listeners and are called to memorize information [18]. The playful nature of robotics is essential in primary education so young students can take the initiative and action [22].

The effect of robotics packages on developing problem-solving skills, the cultivation of thinking following sequences, and students' computational thinking also seem positive. Children learn to argue based on causal relationships [19].

Educational robotics could reduce students' dropping out of school before completing primary education or enhancing the motivation for people belonging to high-risk groups. These groups include women due to gender discrimination or even people with disabilities, people from low socioeconomic status, or even people from diverse cultural backgrounds who, in many cases, may not have the opportunity to engage in innovative activities such as educational robotics. The cultivation of critical and computational thinking, collaboration, creativity, and metacognition through robotics gives students a chance to have more accessible access to their future work environment[1] [10].

Research conducted with preservice and in-service teachers showed that for most teachers, students show interest in building a robot because of the motivation that increases through engaging with it. They also believe that building a robot is more suitable for elementary school students than designing, applying algorithms, and programming. As for the skills that teachers consider useful for students' daily lives, they seem to believe more in children's ability to work in groups and gain everyday experiences and secondarily in their autonomous learning ability. Through robotics, students can develop critical thinking skills when solving a problem and create cognitive models that are considered very important for future employees [2].

Robotics seems to be more related to the field of Technology and Engineering. Besides, a robot is a tool of technology and a product of engineering. At the same time, the relationship between robotics and Mathematics, and Science is not entirely obvious. In this regard, in their pilot survey report, research data show positive results in developing problem-solving skills and understanding mathematical terms, concepts, and phenomena such as motion in the field of Physics [22]. We can add an increase

[1] Researchers propose a model that clarifies the relationship between educational robotics and what they call "inclusive education" [10].

in motivation and a positive attitude towards these areas through students' involvement with targeted robotic activities.

Robotics can interconnect other courses or subjects across curricular and interdisciplinary approaches [4, 5]. Also, robotics is a way for teachers to get involved with STEM courses, and, as it turns out, this involvement is gradually improving. The most exciting thing is that this change observed in this involvement was not related to the knowledge acquired but increased emotional involvement and teachers' interest. In this case, the researchers justify this increase in emotional involvement with teachers' positive attitudes towards robotic activities' practical nature [21].

According to a survey of teachers who have been involved in robotics or have taken part in a competition, robotic activities are more suitable for technology, physics, engineering, mathematics, and less for courses like history [9]. Similar data show that robotics is applied mainly in STEM courses, and robots play a mediator between students and learning objects. Interestingly, most relevant studies of individual and social skills are developed through robotics centered on students older than six [19, 22].

3.2 What Challenges Arise by Using Educational Robotics?

3.2.1 Teacher

In their review, Xia and Zhong [23] point out that studies have highlighted situations where educational robotics did not have the expected positive results for improving learning. Lathifah, Budiyanto, and Yuana [24] refer to teachers' attitudes towards educational robotics in their review. These attitudes cover a wide range ranging from a positive attitude and the belief that robotics is a useful tool for teaching and learning and linking to STEM courses to believe that educational robotics is not suitable for elementary school children. The teachers also present a negative element, the lack of previous experience in subjects such as programming, while the concern is mentioned whether a robot will negatively affect the teacher's role.

The importance of teachers' perceptions of the impact of robotics on primary education and STEM education is explored by Kanhlari and Kiaie [25] due to the lack of research data on this topic. This study's critical point is that the research participants are teachers with little or no robotics experience. Ten out of eleven teachers, while they are aware of robotics' existence, have not used it due to stress, while one stated that although she intended to use it in teaching, she is discouraged and frustrated when applying it in practice.

While the literature highlights the contribution of robotics in promoting STEM courses, in some studies [14, 21], a significant percentage of participants believe that learning mathematical concepts is not facilitated. One explanation is that students' interest in mathematical concepts is difficult to maintain when technology is involved in the learning process, which excites students' motivations.

It should not be overlooked that many teachers have not been systematically involved in programming, Science, or other STEM courses since their studies [21]. This implies difficulties in implementing educational robotics activities in practice and low levels of self-confidence [24]. Teachers' knowledge can be updated with workshops to design and use robotics packages [19].

Leoste and Heidmets [1] report that a teacher, when using an educational robot, feels anxious due to technical difficulties he/she may face (e.g., robot operation problems) as well as the changes he/she is required to implement due to the different way of teaching, which is student-centered and requires different class organization. This is also relevant to students who need to work in groups and manage more information in the problem-solving process.

Kim, Choi, Han, and So [26] express that as the emergence of new technological tools changes the teacher's role, we need to explore and find ways to enhance teachers' skills and adaptability. Thus, they suggest enhancing thinking skills and not focus solely on technical skills. Their study found that engagement with programming and robotics activities helps teachers develop convergent and divergent thinking, computational, creative, and systematic thinking.

The teacher's role differs depending on the robot's role in the learning process. For example, if the robot is used as a teaching tool, the teacher has a mediating role. If a robot is used to teach one or more subjects, the teacher has a more active role as a facilitator [27].

However, in the last twenty years, robotic kits (LEGO Mindstorms NXT, Arduino, Crickets, and others) have been developed and improved to attract students' enthusiasm and foster various skills [5]. Each platform offers advantages that the teacher needs to know in order to choose the right one. This point also highlights teachers' difficulty due to their pre-existing knowledge about educational robotics and programming. In the literature, this, in many cases, increases the stress and feeling of uncertainty for the teacher [1].

It is worth mentioning the European program Teacher Education on Robotics - Enhanced Constructivist Pedagogical Methods (TERECoP), which aims to train teachers in the concept of robotics, how to design and apply robotic activities. The specific program had an educational character and was based on constructivist theories. Another critical element is creating a community of practice by educators who could ease each other using robotic tools [3][2].

3.2.2 Curriculum – Learning Context

A proposal for a curriculum would include the harmonization of teaching methods with current technologies such as robotics. Especially for elementary school, educational robotics could be seen as a tool through which students can see a different perspective of teaching traditional subjects, come into the first contact with basic concepts of programming and problem solving following a sequence of steps, be introduced to the discovery method through trial and error by exploiting the curiosity they already have by nature. Also, students can work together in a small group and recognize the contribution of each member. Thus, they will perceive robotics as a different source of learning [21].

[2] Based on the TERECOP program, some exciting projects of European countries that participated (Greece, France, Romania, Italy, Spain) were presented, but which are addressed to age groups beyond primary school [7].

Above in this paper, we referred to the theories of constructivism and constructionism as fundamental to educational robotics [7, 22]. In this context, the instructor - educator undertakes to find ways for students to have the chance to be actively involved and to build their knowledge. Through interaction with each other and dealing with real-world problems, they build their learning and understanding and personality [5]. In any case, the activities should be harmonized with the students' needs and abilities and should be of scalable difficulty to complete them with the teacher's continuous support [10].

As Alimisis [5] reports, while there is an increase in the demand for robots personally, either for entertainment or education, the same trend is not observed by the various educational institutions based on OECD data. Curricula do not show flexibility in integrating robotics, which is reflected in reducing students' interest in choosing technological areas for their studies.

The integration of robotics systematically is not foreseen in the curricula [5, 27]. More specifically, there is no clear framework in the existing curricula nor a provision for creating a robotics curriculum to design appropriate teaching materials. This shows that robotics is treated more as an extracurricular activity that falls into non-formal learning. In such a vague context, the role of the teacher remains unclear.

In general, it should be noted that in several cases, the involvement of robotics projects is encouraged by the educational authorities, at least to some extent. Also, many teachers increasingly tend to enrich their teaching with robotic activities. It is essential to mention the difficulties teachers face in this process, such as completing a robotics project. Also, the equipment's high cost, even if it is a single class, and practical issues arise, such as mixing pieces and the difficulty of sorting and placing them in the right place in their kit [5].

4 Discussion - Conclusions

Despite rapid advancements, education insists on maintaining an outdated system in the current technological, economic, and social conditions. However, there is an international effort to reform curricula, encouraging innovative teaching models in primary education that promote STEM education and computational thinking as a problem-solving method.

Robotics is perhaps the most appropriate tool for highlighting aspects of STEM training and creating active creators. So, if we want to increase the future employees in STEM fields, a corresponding increase of the school's relevant courses' interdisciplinarity through the curricula is needed [7].

Students need skills (often referred to as 21st-century skills) to help them cope with modern society's demands. Robotics enhances critical thinking, computational thinking, problem-solving, algorithmic thinking, making decisions, creativity, and collaboration. Educational robotics provides opportunities by integrating many different subjects besides Science, technology, engineering, and mathematics, such as art, dance, music, and more, and therefore its integration into formal learning is reasonable.

Educational robotics may be related to STEM fields such as Science, technology, engineering, and mathematics. For this reason, when robotics is evaluated, the results in students' knowledge in the respective fields are evaluated. This means that for a clearer picture of educational robotics and its use in practice, it is necessary to study its application in other fields [10].

From the reviewed articles[3], it becomes clear that many benefits in the learning process in terms of understanding the concepts and principles of the relevant scientific fields, changing attitudes and behavior towards STEM courses, and improving skills. However, physical experience alone is not enough. Besides, a guided exploratory process is needed from an early age and an appropriate curriculum.

Using educational robotics does not mean that the educational problems are solved automatically. This is a handy tool, but attention needs to be paid to the various dimensions that can affect learning quality and prevent all students' useful inclusion in education [10].

According to the articles studied, learning based on the problem-solving, project, and inquiry methodology on the design and operation makes students more active. Open-ended questions increase students' involvement by making the problem under consideration for their own each time. This means that they have the freedom to choose the approach they want to solve the problem by developing critical thinking, collaboration, problem-solving ability, creativity. In such a context, the teacher facilitates rather than gives directions.

Equally important for the age group under consideration are the playful features of robotics. On the one hand, these features create an enjoyable environment [24]; on the other hand contribute to the development of gamification methodology in the classroom [20]. Especially for elementary school students, this aspect of robotics positively affects individual action and motivation [22].

However, what seems to emerge is that students from their engagement with innovative activities such as robotics, learn the scientific method (exploration, experimentation, evaluation, re-evaluation) and ultimately facilitate the construction of their knowledge and understanding [18] and improve their "science literacy" [23].

A variety of strategies are mentioned for learning robotics. Such are exploratory, inquiry-based, collaborative, problem solving, competitive learning. Working in groups of a few members seems to be a more efficient and scalable activity [7, 24]. Also, strategies such as storytelling occasioned by engaging in robotic activity, especially when combined with courses in various fields (e.g., art and engineering), attract the interest of more students [5], and at the same time, language skills are enhanced [8]. However, teachers can develop creative thinking through storytelling [26].

There are several reviews on implementing educational robotics for primary school students [8, 23]. Despite their restrictions, this reflects scientists' and educators' keen interest in exploring an educational tool such as robotics and its potential impact on the learning process to create opportunities instead of inequalities.

[3] Appendix.

Robotics seems to be getting closer to integrating into formal education design. It is considered necessary to prepare teachers before their official integration into the school. Regarding teachers' suspicion for limiting their role, it should be emphasized that robotics complements and facilitate teaching and learning. The necessary decisions and initiatives belong to the teacher [27], which will be assisted by a clear formal framework [4].

5 Future Research

Investigating educational robotics through the selected articles reveals the lack of reports on the connection of robotics with the latest trends in education such as STEAM, where Science, Technology Engineering and Mathematics, interconnect with Art, and the even more recent STREAM education where the importance of literacy development is recognized (Reading and Writing skills).

Also, we found that there is not enough data on the teacher's role in implementing robotic activities. Many teachers will look for theoretical and practical directions to integrate robotics in teaching practice in the future. However, many teachers will look for theoretical and practical directions to integrate robotics in teaching practice in the future expanding its usage in STEAM and even STREAM education. It would be useful to create a guide with a wide range of robotic activities ranging from activities that require basic programming skills to the most complicated programming skills that older students may have acquired and can be used by teachers in the classroom. For a more complete view on the use of educational robotics, it would help to study from the point of view of students and especially the challenges they may face.

Besides, it would be interesting to study any changes that have taken place in the curricula in recent years in different countries, as they could be an example for those who are at a very early stage [28, 29]. Another issue to explore could be the impact of robotics competitions on the learning process. Finally, it would be essential to enrich the research on robotics' direct effect on personal development to be validated scientifically without being just a fad.

6 Limitations

With this study, it is realized that there is a large number of related studies, theoretical and empirical, about educational robotics. These studies were limited, based on subjective criteria. In addition to empirical studies, most of them come from articles published in magazines. This raises the issue of using other publishing types. Another limitation is related to the absence of a meta-analysis of the review. In any case, the reader is invited to consider the above limitations and critically analyze them in future research.

Appendix

See Table 1.

Table 1. Summary of reviewed articles

Paper	Intervention	Major results	Limitations
Álamo et al. (2019) [20]	Survey (15 questions)	Essential elements are collaboration, the problem-solving process, teachers' training, and time in formal education	
Alimisis et al. (2007) [3]	Pilot courses	Training projects are needed. TERECoP is likely to have positive results	
Blancas et al. (2020) [6]	Six session programme (CREA)	Children preferred crafting activities and teamwork but have difficulties with programming	The sample was limited
Chiazzese et al. (2019) [19]	Robotics laboratory	Robotics foster skills. Suggest teachers' training and peer mentoring	Limited sample. Students' performance is not evaluated
Daniela and Lytras (2019) [10]	Model for "inclusive education"	Educational robotics reduce the number of children dropping out of primary education	
Eguchi (2014) [4]	Robotics projects	Robotics can be part of the curriculum	
Faisal et al. (2012) [18]	Robotic activity. Pre and post-lesson assessment	Engaging students improved their understanding, kept them interested, and helped them connect learning with the real world	
Frangou et al. (2008) [7]	Pilot courses for teachers training	There are innovative practices in teacher training in the integration of robotics	
Karypi (2018) [9]	Questionnaire (5 parts, close-ended questions)	Students are motivated for STEM subjects and are more favorable for robotic projects	Lack of investigation on the participants' views and project characteristics

(*continued*)

Table 1. (*continued*)

Paper	Intervention	Major results	Limitations
Khanlari and Kiaie (2015) [25]	Online survey	For non-experienced teachers, robotics is a useful tool, while students being more active	
Kim et al. (2015) [21]	Experiment	Teachers developed active engagement and positive emotions for robotics and STEM	Limited internal validity and reliability
Kim et al. (2012) [26]	Three cases	New technologies are a means of developing skills, broadening teachers' perceptions	Limited generalizability
Lathifah et al. (2019) [24]	Review of literature	Not all educators have a positive attitude towards the use of new technologies	A limited number of papers
Leoste and Heidmets (2020) [1]	Educational material was developed	Most teachers need training. Applying robotics requires more time and occupation	Small sample and short duration
Mubin et al. (2013) [27]	Review of literature	Robots do not replace the teacher	The omission of studies
Ospennikova et al. (2015) [11]	Pedagogical experiment	Robotics can be integrated into formal education. The percentage of children who choose physics or math in high school is growing	
Roussou and Rangoussi (2020) [12]	Applied experiment	Using robots pedagogically and playfully improves students' skills	Limited generalization of results
Smyrnova-Trybulska et al. (2016) [2]	Survey (15 questions)	Creating laboratories is an essential form of an interdisciplinary approach	
Stergiopoulou et al. (2016) [22]	Experiment	Students seemed to understand the interconnection of STEM courses	Small sample

(*continued*)

Table 1. *(continued)*

Paper	Intervention	Major results	Limitations
Toh et al. (2016) [8]	Systematic review	Robotics develops skills, increases motivation and interaction	
Xia and Zhong (2018) [23]	Systematic review	Most studies were non-experimental. In some cases, the results did not show improvement	

References

1. Leoste, J., Heidmets, M.: Bringing an educational robot into a basic education math lesson. In: Merdan, M., Lepuschitz, W., Koppensteiner, G., Balogh, R., Obdržálek, D. (eds.) Robotics in Education. RiE 2019. Advances in Intelligent Systems and Computing, vol. 1023. Springer, Cham (2019). http://doi-org-443.webvpn.fjmu.edu.cn/10.1007/978-3-030-26945-6_21
2. Smyrnova-Trybulska, E., Morze, N., Kommers, P., Zuziak, W., Gladun, M.: Educational robots in primary school teachers' and students' opinion about STEM education for young learners. In: International Association for Development of the Information Society (2016)
3. Alimisis, D., Moro, M., Arlegui, J., Pina, A., Frangou, S., Papanikolaou, K.: Robotics and constructivism in education: the TERECoP project. In: Kalas, I. (ed.) Proceedings of the 11th European Logo Conference, Bratislava, Comenius University (2007). ISBN 978-80-89186-20-4
4. Eguchi, A.: Robotics as a learning tool for educational transformation. In: Proceedings of 4th International Workshop Teaching Robotics, Teaching with Robotics and 5th International Conference Robotics in Education, Padova (2014)
5. Alimisis, D.: Educational robotics: open questions and new challenges. Themes Sci. Technol. Educ. **6**(1), 63–71 (2013)
6. Blancas, M., Valero, C., Mura, A., Vouloutsi, V., Verschure, P.: "CREA": an inquiry-based methodology to teach robotics to children. In: 10th International Conference on Robotics in Education, Vienna (2019)
7. Frangou, S., Papanikolaou, K., Aravecchia, L., Montel, L., Ionita, S., Arlegui, J., Pina, A., Menegatti, E., Moro, M., Fava, N., Monfalcon, S., Pagello, I.: Representative examples of implementing educational robotics in school based on the constructivist approach. In: Proceedings of International Conference on Simulation, Modeling and Programming for Autonomous Robots, pp. 54–65, Venice (2008)
8. Toh, L.P.E., Causo, A., Tzuo, P.W., Chen, I.M., Yeo, S.H.: A review on the use of robots in education and young children. Educ. Technol. Soc. **19**(2), 148–163 (2016)
9. Karypi, S.: Educational robotics application in primary and secondary education: a challenge for the Greek teachers society. J. Contemp. Educ. Theory Res. **2**(1), 9–14 (2018)
10. Daniela, L., Lytras, M.D.: Educational robotics for inclusive education. Tech. Know. Learn. **24**, 219–225 (2019). https://doi.org/10.1007/s10758-018-9397-5
11. Ospennikova, E., Ershov, M., Iljin, I.: Educational robotics as an innovative educational technology. Social and Behavioral Sciences **214**, 18–26 (2015). https://doi.org/10.1016/j.sbspro.2015.11.588

12. Roussou, E., Rangoussi, M.: On the use of robotics for the development of computational thinking in kindergarten: educational intervention and evaluation. In: Merdan, M., Lepuschitz, W., Koppensteiner, G., Balogh, R., Obdržálek, D. (eds.) Robotics in Education. RiE 2019. Advances in Intelligent Systems and Computing, vol. 1023. Springer, Cham (2020). http://doi-org-443.webvpn.fjmu.edu.cn/10.1007/978-3-030-26945-6_3
13. Deeks, J.J.: Systematic reviews of published evidence: miracles or minefields? Ann. Oncol. **9**(7), 703–709 (1998). https://doi.org/10.1023/a:1008335706631. PMID: 9739434
14. Lewis, G., Churchill, R., Hotopf, M.: EDITORIAL: systematic reviews and meta-analysis. Psychol. Med. **27**(1), 3–7 (1997). https://doi.org/10.1017/S0033291796004291
15. Moher, D., Liberati, A., Tetzlaff, J., Altman, D.: Preferred reporting items for systematic reviews and meta-analyses: the PRISMA statement. J. Chinese Integrative Med. **7**(9), 889–896 (2009). https://doi.org/10.3736/jcim20090918
16. Pai, M., McCulloch, M., Gorman, J.D., Pai, N., Enanoria, W., Kennedy, G., Tharyan, P., Colford Jr., J.M.: Systematic reviews and meta-analyses: an illustrated, step-by-step guide. Nat. Med. J. India **17**(2), 86–95 (2004). PMID: 15141602
17. Tawfik, G.M., Dila, K.A.S., Mohamed, M.Y.F., Tam, D.N.H., Kien, N.D., Ahmed, A.M., Huy, N.T.: A step by step guide for conducting a systematic review and meta-analysis with simulation data. Trop. Med. Health **47**, 46 (2019). https://doi.org/10.1186/s41182-019-0165-6
18. Faisal, A., Kapila, V., Iskander, M.G.: Using robotics to promote learning in elementary grades. In: ASEE Annual Conference and Exposition, San Antonio (2012). https://doi.org/10.18260/1-2-22196
19. Chiazzese, G., Arrigo, M., Chifari, A., Lonati, V., Tosto, C.: Educational robotics in primary school: measuring the development of computational thinking skills with the bebras tasks. Informatics **6**(4), 43 (2019). https://doi.org/10.3390/informatics6040043
20. Álamo, J., Quevedo, E., Marqués, J.P.: Integration of educational robotics with active didactic methodologies in primary school. In: VI Jornadas Iberoamericanas de Innovación Educativa en el Ámbito de las TIC y las TAC, Las Palmas de Gran Canaria (2019)
21. Kim, C., Kim, D., Yuan, J., Hill, R.B., Doshi, P., Thai, C.N.: Robotics to promote elementary education preservice teachers' STEM engagement, learning, and teaching. Comput. Educ. **91**, 14–31 (2015). https://doi.org/10.1016/j.compedu.2015.08.005
22. Stergiopoulou, M., Karatrantou, A., Panagiotakopoulos, C.T.: Educational robotics and stem education in primary education: a pilot study using the H&S electronic systems platform. In: International Conference EduRobotics (2016). https://doi.org/10.1007/978-3-319-55553-9_7
23. Xia, L., Zhong, B.: A systematic review on teaching and learning robotics content knowledge in K-12. Comput. Educ. **127**, 267–282 (2018)
24. Lathifah, A., Budiyanto, C.W., Yuana, R.A.: The contribution of robotics education in primary schools: teaching and learning. In: Proceedings of AIP Conference, vol. 2194, p. 020053 (2019). https://doi.org/10.1063/1.5139785
25. Khanlari, A., Kiaie, F.M.: Using robotics for STEM education in primary/elementary schools: teachers' perceptions. In: 10th International Conference on Computer Science and Education (ICCSE 2015), Fitzwilliam College, Cambridge University (2015)
26. Kim, H., Choi, H., Han, J., So, H.J.: Enhancing teachers' ICT capacity for the 21st-century learning environment: three cases of teacher education in Korea. Aus. J. Educ. Technol. **28**(6) (2012). https://doi.org/10.14742/ajet.805

27. Mubin, O., Stevens, C., Shahid, S., Al Mahmud, A., Dong, J.J.: A review of the applicability of robots in education. Technol. Educ. Learn. **1**, 1–7 (2013). https://doi.org/10.2316/journal.209.2013.1.209-0015
28. Papadakis, S.: Robots and robotics kits for early childhood and first school age. Int. J. Interactive Mobile Technol. (IJIM) **14**(18), 34–56 (2020). https://doi.org/10.3991/ijim.v14i18.16631
29. Kalogiannakis, M., Papadakis, S., Zourmpakis, A.-I.: Gamification in science education. A systematic review of the literature. Educ. Sci. **11**(1), 22 (2021)

Author Index

M. Malvezzi et al. (Eds.): EDUROBOTICS 2021, SCI 982, pp. 231–232, 2021.
https://doi.org/10.1007/978-3-030-77022-8